A Game-Theoretic Perspective on Coalition Formation

The Lipsey Lectures

The Lipsey Lectures, delivered every two years, offer a forum for leading scholars to reflect upon their research. Lipsey lecturers, chosen from among professsional economists approaching the height of their careers, will have recently made key contributions at the frontier of any field of theoretical or applied economics. The emphasis is on novelty, originality and relevance to an understanding of the modern world. It is expected, therefore, that each volume in the series will become a core source for graduate students and an inspiration for further research.

The lecture series is named after Richard G. Lipsey, the founding professor of economics at the University of Essex. Professor Lipsey instilled at Essex a commitment to explore challenging issues in applied economics, grounded in formal economic theory, the predictions of which were to be subjected to rigorous testing, thereby illuminating important policy debates. This approach remains central to economic reasearch at Essex and an inspiration for members of the Department of Economics. In recognition of Richard Lipsey's early vision for the Department, and in continued pursuit of its mission of academic excellence, the Department of Economics is pleased to organize the lecture series, with support from Oxford University Press.

A Game-Theoretic Perspective on Coalition Formation

DEBRAJ RAY

OXFORD

UNIVERSITY PRESS

OXFORD

UNIVERSITY PRESS

Great Clarendon Street, Oxford OX2 6DP

Oxford University Press is a department of the University of Oxford.
It furthers the University's objective of excellence in research, scholarship,
and education by publishing worldwide in

Oxford New York

Auckland CapeTown Dar es Salaam Hong Kong Karachi
Kuala Lumpur Madrid Melbourne Mexico City Nairobi
New Delhi Shanghai Taipei Toronto

With offices in

Argentina Austria Brazil Chile Czech Republic France Greece
Guatemala Hungary Italy Japan Poland Portugal Singapore
South Korea Switzerland Thailand Turkey Ukraine Vietnam

Oxford is a registered trade mark of Oxford University Press
in the UK and in certain other countries

Published in the United States
by Oxford University Press Inc., New York

© Debraj Ray 2007

British Library Cataloguing in Publication Data

Data available

Library of Congress Cataloging in Publication Data

Data available

Typeset by Author Using LaTeX
Printed in Great Britain
on acid-free paper by
Biddles Ltd., King's Lynn, Norfolk

ISBN 978-0-19-920795-4

1 3 5 7 9 10 8 6 4

To Rahul Wahi

(1956–2006)

Preface

This monograph describes a theory of coalition formation. I would like it to serve as an open invitation for young theorists to enter this fascinating and important line of inquiry. My own account of this theory is naturally selective and largely based on research that I have conducted, so I cannot hope to have been comprehensive in any way. I am uneasily aware, for instance, of the vast stylistic and expositional differences between this work and my earlier book, *Development Economics*. Individuals seeking the same degree of coverage will certainly not find it here. Yet I hope that in some way this short book will have enough in it to attract, provoke, and even be occasionally useful.

The basic objective of this book is easy enough to describe. I outline a theory of *coalition formation*, a process by which individual agents come together to achieve collaborative though occasionally compromised goals. The underlying premise of the theory is simple yet compelling. Left to their own devices, individuals will generally engage in actions that fail to adequately internalize the negative externalities imposed on one another. The explicit agreement to form a coalition may be viewed as an agreement to be jointly sensitive to those externalities, and to take actions to try and lower them.

A theory of coalition formation that starts from this premise can proceed along one of two broad lines. First, the grouping of individuals into coalitions may simply represent a certain degree of consensus, with no binding agreements involved. In such a line of inquiry, the principal focus of attention would be the actual strategies that sustain such agreements, as well as the "best" agreements sustainable in nonbinding play. A leading framework for such a study is the theory of dynamic games (especially repeated games).

A second view, central to the tradition of cooperative game theory, is that an agreement, once made, is binding. Just *how* it is binding is not really up for discussion. It may be that there are enough strategic checks and balances (in repeated play, for instance) to keep the agreement together. Or perhaps social conventions and the threat of social sanctions uphold an agreement. Or perhaps an agreement can be legally enforced in ways that — while possibly fascinating to a lawyer — are not of great interest to the game theorist. The focus is on the agreements themselves, and especially on the process by which those agreements are *reached*, as opposed to the way in which those agreements are *implemented*.

Obviously, the two views are complementary. They study two distinct aspects of the theory of coalition formation.

This book is a contribution to the second view. In this sense, then, my questions belong to the old-fashioned, classical lineage of cooperative game theory. But the approach is very different. While agreements once written are presumed to be binding, the *process* of achieving those negotiations is assuredly not. Proposal and counterproposal, acceptance and rejection, objection and agreement, sidepayments and subcoalitions, all play a role in the strategic negotiations that precede the writing of agreements. Thus much of the material in this book represents a marriage of noncooperative game theory and its more traditional, cooperative counterpart, and I hope — as the title suggests — that it will provide a useful game-theoretic perspective on coalition formation.

As we shall see, the process of arriving at an agreement is far from trouble-free, even in a world of perfect and complete information. While there are clear incentives for all parties to come together in the interests of minimizing externalities and thereby achieving efficiency, there is also room for subterfuge and sabotage, for the formation of intermediate coalitions that may profit from an inefficient situation. These possibilities influence a small set of questions that run through the book: What agreements will be written? Which coalitions will form? Are binding agreements invariably efficient?

This book is a substantial outgrowth of the inaugural Richard Lipsey lectures given at the University of Essex in December 2004. I have tried to provide a self-contained and rigorous account, emphasizing the conceptual issues involved but without skimping on the necessary formalism. With interest in economic theory at an all-time low (for good reason, some would say), I don't expect the battle-scarred veteran economist to give the more difficult arguments more than a cursory once-over. But I do hope that the younger, more open, non-bottom-line oriented generation of economic theorists will read some of this material with interest, with an eye to taking the story much further, and in many more imaginative directions than I could ever expect to address.

I am immensely grateful to Abhinay Muthoo, who — as Department Chair at the University of Essex — first invited me to give these lectures, and then (before I could react swiftly enough) talked me into writing this book. Thank you Abhinay, I would have never done it otherwise. I am also very grateful to Venkataraman Bhaskar and Sanjeev Goyal, now ex-Essex but certainly leading figures in the Economics Department when they were there, who offered constant encouragement and support when I dithered with the choice of subject matter for the Lectures. Yes, you *can* pull off a public lecture on game theory, they said. I hope they were right.

My greatest intellectual debt is to Rajiv Vohra. He and I have been thinking together about the issues in this book since 1988. Half the book is based on my joint work with him and while he graciously encouraged me to write this monograph on my own, it is only right that I acknowledge, here in unambiguous print, that I view him as possessing the moral equivalent of full coauthorship rights (and obligations in case of any errors).

I want to especially mention three other influences on this book: Kalyan Chatterjee, Bhaskar Dutta and Kunal Sengupta. Together we wrote a four-author paper that I am proud of and that served as foundation — implicit or explicit — for many of the ideas presented here. It was a great intellectual experience (and better yet, fun) to work on these ideas together at the Indian Statistical Institute in the 1980s, where Bhaskar, Kunal and I were at the time, with Kalyan a regular visitor.

Two other coauthors figure prominently in this book: Hideo Konishi and Kyle Hyndman. I met Hideo when he visited Boston University in 1997. It didn't take long to see that we had several interests in common, and we worked together on a paper that makes an appearance in this book. Kyle Hyndman (as a graduate student at NYU) and I worked on another paper that plays a role in this book. I'm very grateful to both Kyle and Hideo for allowing me to use their research.

I want to record my gratitude to some of the other individuals whose research has (indirectly or directly) shaped my own interests. My somewhat narcissistic focus notwithstanding, several of them are mentioned in this book. It couldn't be otherwise. I especially have in mind Francis Bloch, Doug Bernheim, Joan Esteban, Armando Gomes, Peter Hammond, Matt Jackson, William Lucas, Roger Myerson, Ariel Rubinstein, David Schmeidler, Lloyd Shapley, Robert Wilson and John von Neumann. (I haven't met all these individuals, at least one of them for demographic reasons, but that doesn't matter.)

I thank Anja Sautmann for her careful reading of the manuscript and suggestions for better exposition. I am grateful to the London School of Economics, STICERD in particular, for hosting me during a wonderful sabbatical year in which the writing of this book began. I thank Luis Cabral and the Stern School at New York University for providing me with a hideaway: an office to retreat to once in a while to think about this book. And I warmly thank Sarah Caro and Carol Bestley at Oxford University Press: Sarah for her encouragement and good cheer through the inevitable delays in writing, Carol for her help with typesetting. While on this last subject, salaams are due to the great Donald Knuth. How would one write this book (or *any* book) without TeX?

No words suffice to thank my wife Nilita and my two children, Riyaaz and Zayira. All I can say is I've never believed you can write a good book when you're happy. I hope I'm wrong.

Contents

CHAPTER 1

Introduction

A group of agents get together to write one or more agreements. Each agent controls some set of actions. If a subset of the agents (possibly the entire set of agents) forms a *coalition*, this means that they agree to behave "cooperatively", to choose and implement a joint course of action. If more than one coalition forms, then players across such coalitions do not proceed jointly; their actions are chosen independently and noncooperatively. We are interested in the equilibrium of coalition formation. Which coalitions will come about? Is there a presumption that such coalition formation will be efficient? If so, under what circumstances? If not, can one place bounds on the extent of inefficiency that may occur?

A variety of situations fits this description well. Take, for instance, the formation of customs unions. A union will cooperate on the relevant actions: tariff setting, the imposition or removal of intra-union boundaries on input flows, or the taking of retaliatory (or indeed provocative) action. *Across* unions, in contrast, there is noncooperative play.

Or consider cartel formation among some firms in an industry. The cartel members will cooperate in output or price-setting, or in the setting of other relevant parameters, such as quality. Across firms in different cartels all such bets are off.

Or consider the signing of environmental agreements across regions or countries. Again, within an environmental coalition members presumably agree to abide by emission protocols, while across coalitions there is no such explicit promise.

I could go on. Protectionism, conflict, strategic voting, R&D agreements, reciprocity, the formation of political parties: these situations and many more have the joint cooperate/compete feature that I emphasize here. In some of these situations a *single* coalition — the *grand coalition* — may well form, and there is full cooperation. In others no coalitions may form, so that agents stay on their own. But these outcomes must be viewed in their ambient context: that there is inevitably potential for various mixes of cooperation and competition. The analysis of this mixture — and the outcomes it generates — is the subject of this monograph.

I don't have an infinite amount of space — or even a great deal of it — so I am going to explicitly lay down two central restrictions at the very outset. I am going to study situations in which there is *no* incompleteness of information, and in which the agreement to cooperate can be made binding at no cost. These restrictions take me squarely into the arena of classical cooperative game theory, but for reasons that I shall presently explain, I will need to go significantly beyond that paradigm.

Do I believe in these restrictions? I certainly do not, in that I recognize the existence of an entire gamut of questions based either on the presumption that binding agreements *cannot* be written, or on a chronic failure of complete information. An enormous literature in game theory is based on one or the other (or both) of these two failures, and I have nothing to add to that literature here. At the same time, it seems to me that there is a full range of issues — including the ones I've mentioned above — for which the question of incomplete information is secondary, and for which it isn't a bad thing to presume that an agreement will be honored. This monograph is based on the premise that there is much to be explored in such a situation, and that existing theory falls short in achieving such an exploration. (I shall soon explain why.)

The questions I ask are certainly not invented by me. The issues of coalition formation and negotiation (in a context where agreements can be made binding), are central to the theory of strategic behavior, and were first explored under the umbrella of "cooperative game theory". That theory presumes that binding agreements can indeed be costlessly written, and seeks to understand the agreements that will, indeed, come about. The Nash bargaining solution, the core, the Shapley value or the stable set of von Neumann and Morgenstern are only a few of the many different solution concepts that have been advanced to answer the question of equilibrium agreements. Some of the greatest minds in the subject — John von Neumann and Oskar Morgenstern, John Nash, Ariel Rubinstein, Lloyd Shapley, Robert Aumann and Roger Myerson, to name some of them in no particular order — have been concerned with the question of describing cooperative agreements. It is certainly possible that the *process* of arriving at such agreements may be noncooperative, but nonetheless it was no problem for any of these authors to visualize situations in which an agreement, *once made*, could be costlessly implemented. Yet the subsequent literature has decidedly moved away from cooperative game theory. It is interesting to speculate on the decline of that theory. There are several possible reasons for this, but I will highlight three.

First, there has been great success with the development of solution concepts for *non*binding agreements, at the heart of which lies the Nash equilibrium. The beauty and utility of the Nash equilibrium concept has lured — and quite understandably so — a great many social scientists into studying its applications. At a more theoretical level, Nash equilibrium has been stretched in both directions, with notions such as rationalizability broadening the concept, and others, such as subgame perfection or sequential equilibrium serving to refine it. It may still be fashionable to pay lip service to von Neumann and Morgenstern's monumental *Theory of Games and Economic Behavior*, but much (indeed, most) of that book is currently in disuse as far as modern, "mainstream" game theory is concerned.

Second, it appears that Ronald Coase's work has, at least to some extent, pushed the profession away from the study of binding agreements, or at least the study of such agreements under conditions of complete information. Provided that all available information is at hand to all parties and that binding agreements can be written, Coase (1960) asserted that equilibrium bargains *must* be efficient. In fact, if payoffs can be freely transferred across agents, then such efficiency must typically imply a determinate course of action independent of bargaining protocol or the distribution of bargaining power, these characteristics only determining how the resulting surplus from the efficient arrangement is divided among the agents. To be sure, if no sidepayments are possible, then the particular outcome will generally vary with bargaining power, but nevertheless it will still be Pareto-efficient. This is the well-known "Coase theorem".

The Coase theorem implies that if you must look for strategic failures of efficiency, whether in markets or in other institutional settings, you must study situations in which information is less than complete, or binding agreements cannot be written. Of course, such settings are important and widespread in the social sciences, and the literature studying inefficient outcomes in such contexts is enormous.

Yet there appear to be numerous situations in which agreements *could* be written, and the relevant information is at hand to all parties, yet the division of society into opposed coalitions, and the consequent inefficiencies that such divisions entail, are endemic. Of course, it is logically possible to maintain the somewhat stubborn position that in all such situations there *must* be some incompleteness of information or some dimension along which a binding agreement cannot be written. But there is another view, and this is what I wish to explore: that while an agreement, *once reached*, can be costlessly implemented, the *negotiation process* leading up to it must be fundamentally modeled as a noncooperative game. Moreover, there is good reason to believe that in many situations, the outcome of that noncooperative game may be inefficient, and so, therefore, are the agreements that are finally implemented.

Even though I do not claim widespread applicability and ultimate definitiveness for the particular analysis in this book, this point of view is important.

Finally, as Chapter 2 and the rest of this book will clarify at various points, the *way* in which cooperative game theory has traditionally developed makes it difficult (and occasionally impossible) to apply that theory to many situations of serious strategic interest. For instance, the *characteristic function* is a fundamental device in the theory of cooperative games. The characteristic function essentially assigns a value, or worth, to every coalition of agents in a game, and this is the starting point for various solution concepts. As I argue in Chapter 2, this is a perfectly reasonable shorthand for a variety of economic and political situations, but it may be an awkward shorthand for those situations in which cross-player externalities are of fundamental significance in the very determination of what coalitional "worth" means. Unfortunately, many — perhaps most — situations of interest to the game theorist falls into this category. Therefore cooperative game theory as it has been traditionally developed fails to answer its questions across a broad enough canvas, important though those questions may be.

A good part of my task in this monograph will be to reformulate and redevelop the cooperative approach in a more general and applicable context. To do so, I will borrow ideas from both cooperative and noncooperative game theory. Far from suggesting that the ideas here represent a radical break from these theories, I extend (and often merge) the two theories in developing the analysis described here.

Part 1

The Setting

The Secrets

CHAPTER 2

Ingredients for a Theory of Agreements

2.1 Introduction

A central question in cooperative game theory is to describe the set of allocations which agents in the game might jointly agree to, and the set of agreeing groups or coalitions that arise as a result. Hence the appellation "cooperative". At least in a one-shot static context, the ground rules are simple: any agreed-upon allocation among a subset of players (including, perhaps, the grand coalition of all players) can be made binding at little or no cost. The mechanics of writing such a binding agreement take a back seat in these proceedings. Indeed, this presumed ability to write binding agreements at will, without worrying about *how* such agreements are to be enforced, is what distinguishes cooperative game theory from its better-known counterpart, the theory of noncooperative games.

The project sounds innocuous enough, but of course there is the small matter of "agreeing upon" the allocation. One might leave such matters to the players themselves, as Ronald Coase effectively did, arguing that somehow or the other the players in question must alight upon an efficient allocation. Or — following the pioneering efforts of John von Neumann and Oscar Morgenstern — one might be interested in analyzing the negotiation process itself. That is the subject of this monograph.

In this chapter, we introduce the different ingredients that go into the theory to be developed. First, we introduce and then significantly broaden a central concept in cooperative game theory: the *characteristic function*. We then describe two ways of thinking about negotiations and agreements, one based on noncooperative theories of bargaining, the other on the notion of blocking coalitions. Next, we discuss the need for theories based on farsightedness. Section 2.5 illustrates these points through the use of two examples. We end the chapter by discussing various assumptions regarding commitment ability, and distinguish in particular between models based on irreversible commitments and theories of binding agreements based on ongoing renegotiation.

2.2 Characteristic Functions and Cooperative Games

The central object over which negotiations occur is easy enough to write down. It is our usual notion of a game in normal or strategic form, with the difference that players are able to implement binding agreements. Formally, let N be a set of players, A_i the set of actions for player i with product set $\mathbf{A}$, and $u_i : \mathbf{A} \to \mathbb{R}$ player i's payoff function. We seek a profile of actions $\mathbf{a} = (a_1, \ldots, a_n)$ that players would "willingly" be signatories to.

Oddly enough, a visitor to the land of cooperative game theory is unlikely to see this starting point. Instead, she would most likely be introduced — in the words of Shubik (1983) — to the "cornerstone of the theory of cooperative n-person games," the *characteristic function*. Continuing to quote Shubik, "[t]he idea is to capture in a single numerical index the potential *worth* of each coalition of players. The characteristic function is, in a sense, the final distillation of the descriptive phase of the theory" (1983, p. 128).

The idea behind the characteristic function is extremely simple. Attach to each nonempty subset of players, or coalition, a set of possible payoff vectors for the players in that coalition. Formally, for each nonempty $S \subseteq N$, a characteristic function $\mathbf{U}$ attaches a set of $|S|$-dimensional vectors $\mathbf{U}(S)$. The Shubik description above actually refers to the special case of a *transferable utility* (TU) characteristic function. Such a function attaches to each coalition S a number $v(S)$, describing the overall worth of that coalition. $\mathbf{U}(S)$ is then the set of all divisions of that worth among the players in S. Our more general formulation handles nontransferability or limited transfers, broadly referred to as the NTU case.

How is a game in strategic form "converted" to a characteristic function? The "standard approach" pioneered by von Neumann and Morgenstern (see also Aumann (1961) and Scarf (1971)) attempts to define payoffs available to each coalition by means of a simplifying heuristic. For instance, the *α-characteristic* function is generated from the original game by allowing a payoff vector **v** to lie in **U**(S) if and only if S has a joint strategy that *guarantees* its players at least **v**, *no matter what players elsewhere do.*[1]

The great virtue of this conversion is its simplicity. It makes the worth of a coalition independent of the ambient environment. Moreover, in many cases it represents a perfectly sensible first cut. The characteristic function works well when externalities across coalitions can be safely ignored.

When externalities are salient, however, the characteristic function is an odd device to say the least. The function is essentially constructed on the presumption that a coalition does not expect to receive more than what it does when outsiders act to sabotage this coalition as best as they can. Of course, there is no reason why the outsiders should behave in this bloodthirsty fashion, and there is no reason for the deviating coalition to *necessarily* expect or fear such behavior.[2]

Consider, for instance, a static Cournot duopoly. There are three coalitions: the grand coalition and the two singletons. What payoff can a single firm obtain under the derived α-characteristic function? Well, if its rival can produce enough output to drive price down to cost (typically assumed in all textbook examples), then the answer is simple: zero or less. But this is clearly absurd: by breaking off negotiations, a firm should surely be able to look forward to Cournot–Nash payoffs in the resulting noncooperative environment.

Matters are, of course, far more complicated when there are more than two players and nonsingleton subcoalitions might form. A principal objective of the theories we develop is to handle such cases, but the main point in the present discussion is simply to show that simplistic conversions to characteristic functions are misleading or sometimes plain wrong. It is even possible that the use of the characteristic function among cooperative game theorists

---

[1]Another way of making the conversion is to define a *β-characteristic* function; see, e.g., Shubik (1983, p. 136–138) for a discussion. The criticism we develop in the main text applies with equal force to this conversion as well.

[2]Neither can one justify this on the grounds of finding a conservative solution. Making things more conservative for a coalition will make things easier for another coalition to which this coalition might object, and this zigzag of alternating conservatism and expansiveness will typically echo its way through chains of objections and counterobjections.

is the single most important cause of its neglect elsewhere in the profession.

While I can't speculate on just why von Neumann and Morgenstern adopted this device, a perusal of their monumental *Theory of Games and Economic Behavior* (1944) yields interesting insights. After their justly celebrated foray into decision theory under risk, and some general remarks on games, they embark on the topic of zero-sum (or equivalently, constant-sum games) on p. 85. This journey continues till p. 504. *Thus more than 400 pages of their 740-page book is devoted to zero-sum games.*

Is it merely a coincidence, then, that the device of reducing a strategic game to a characteristic function makes far better sense in a zero-sum game? (My coalition's loss is your complementary coalition's gain, so it is reasonable for your coalition to minimax mine, and for mine to behave on that presumption.) The question is rhetorical. Of course it isn't a coincidence. Indeed, it is amply clear that by the time von Neumann and Morgenstern take up general games on p. 504, they are too enamoured of the characteristic function to let go.

Of course, this isn't to say that they ignore the problem. They take it up explicitly on p. 540:

> The desire of the [complementary] coalition $-S$ to harm its opponent, the coalition S, is by no means obvious. Indeed, the natural wish of the coalition $-S$ should not be so much to decrease the expectation value . . . of the coalition S as to increase its own expectation value. These two principles would be identical . . . when Γ is a zero-sum game, but it need not be at all so for a general game . . .
>
> I.e. in a general game . . . the advantage of one group of players need not be synonymous with the disadvantage of the others . . . In other words, there may exist an opportunity for genuine increase of productivity, simultaneously in all sectors of society.
>
> . . . Indeed, this is more than a mere possibility — the situations to which it refers constitute one of the major subjects with which economic and social theory must deal. Hence the question arises: Does our approach not disregard this aspect altogether? Did we not lose this cooperative side of social relationships because of

the great emphasis which we placed on their opposite, antagonistic side?

Having demonstrated that they are very aware of the limitations, they go ahead anyway. Two defences are mounted, neither very satisfactory. First, they argue that "inflicting losses on the adversary may not be directly profitable in a general game, but it is the way to put pressure on him. He may be induced by such threats to pay a compensation, to adjust his strategy in a desired way, etc." (p. 541). This is obviously not convincing at all, or at the very least would require folk-theorem-like arguments in a dynamic setting, which then puts us into a different strategic form game altogether. The authors aren't convinced either: "It must be admitted, however, that this is not a justification of our procedure — it merely prepares the ground for the real justification which consists of success in examples" (p. 541).

Their second defence is that all games may be viewed as constant-sum anyway, provided that sidepayments can be made in a trans-ferable manner. This Coaseian line of reasoning has its flaws, however. Once a coalition moves off to be on its own, one cannot *continue* to assume that efficient sidepayments will be made across coalitions. Once again staying within the one-shot context, all unilateral payments are part of a system of binding agreements: if these are indeed being made to or from coalition S, then S isn't a coalition on its own! We haven't yet defined what it means for a coalition to be on its own, but whatever that might mean, it must imply *nonbinding* play between that coalition and the "outside world".

Von Neumann and Morgenstern conclude their discussion thus: "In spite of all this, the reader may feel that we have overemphasized the role of threats, compensations, etc., and that this may be a one-sidedness of our approach which is likely to vitiate the results in applications. The best answer to this is ... the examination of those applications".

Indeed, we will examine some applications in this monograph. Nevertheless, it is best to state our final view as quickly as possible. There are situations in which characteristic functions may be useful: those in which there are *genuinely* no externalities across players, or when the sum of payoffs is constant. Indeed, many such situations come to mind: the provision of *local* public goods, voting games, bargaining, or endowment-trading games with no external effects. We shall also see that a consideration of characteristic functions provides useful methodological steps towards a more general analysis. But having stated these points, it must also be asserted clearly that characteristic functions must be dispensed with in a genuine externality-ridden world, especially if the theory of cooperative games is to have any lasting effect on the profession at large.[3]

2.3 Two Approaches to Coalition Formation

The next ingredient concerns methodology. This book studies two approaches to coalition formation. The first approach is close to the "standard" methodology of game theory in that the entire process of coalition formation and writing binding agreements is itself modeled as a traditional noncooperative game. Protocols exist for individual proposals and individual responses, and every proposal is made by, or accepted or rejected, by an individual.

Part 2 of the book epitomizes this approach. While agreements may be fully cooperative and binding *provided* all parties agree, the process of agreement may itself be modeled as a noncooperative game. Such a game is described by a process that selects *proposers* and *responders* according to a *protocol*. A proposer attempts to form a coalition — perhaps the grand coalition of "full cooperation" — by making a proposal for that coalition to form. Attention then shifts to the members of that coalition, who must either accept or reject the proposal. Acceptance means that the coalition in question has formed, and the negotiation game continues with the remaining set of agents. Rejection means that the very same set of players continues into the next round, and the protocol selects a fresh proposer.

In Part 3 of the book we take up a second approach, one more in line with the traditional theory of cooperative games. Under this view, *coalitions* are treated as fundamental units. While proposals may or may not be made by individuals or groups, attention is deliberately not placed on how the proposals come about. Responses to such proposals are firmly grounded at the coalitional level, but not through

---

[3]Early attempts to extend the idea of characteristic functions to incorporate externalities include Lucas (1963), Thrall and Lucas (1963), and Rosenthal (1972).

an explicit game form based on *individual* moves. Objections will, instead be studied as coalitional *blocks*.

The two approaches are based on distinct philosophical foundations. The "blocking approach" that we borrow and adapt from classical cooperative game theory treats coalitions as fundamental behavioral units. That isn't to say that individuals don't matter — they certainly do — but there isn't an explicit decision-theoretic foundation to these models, based firmly on methodological individualism. In contrast, what we shall call the "bargaining approach" to coalition formation has strong decision-theoretic foundations, grounded at the individual level.

2.4 Farsightedness

Central in all we do is the notion of *farsightedness*. Whether based on blocking or bargaining, a theory of group formation must come to grips with the possibility that agents look beyond the immediate consequence of their own actions. In their paper on the strategic formation of bilateral links, Aumann and Myerson (1988) observe:

> When a player considers forming a link with another one, he does not simply ask himself whether he may expect to be better off with this link than without it, given the previously existing structure. Rather, he looks ahead and asks himself, 'Suppose we form this new link, will other players be motivated to form further new links that were not worthwhile for them before? Where will it all lead? Is the *end result* good or bad for me?'

Similarly, in any theory of group formation, a player or group of players breaking off negotiations must do more than simply presume that they will be engaged in a noncooperative game with the resulting complementary coalition. They must be prepared for two kinds of possible repercussions:

[1] The "deviant group" must be aware that (until they have signed some agreement of their own) they are potentially vulnerable to *further* deviations by members of their own group.

[2] In addition, the group must attempt to predict the coalition structure that arises elsewhere and not just presume that the larger group they have broken from will simply band together.

It is worth noting that the considerations raised in [1] are similar to issues that arise in coalitional refinements of *nonbinding* equilibrium play. Indeed, the notion of a coalition-proof Nash equilibrium (introduced in Bernheim, Peleg and Whinston (1987)) is based squarely on a formalization of [1], so that potential coalitional deviations are tempered by the realization that such deviations may be "susceptible" to "further" deviations.[4] Parallel considerations apply to the notion of blocking in cooperative games: might a block not be subjected to further blocks?[5]

But item [2] is distinctive, in that it generally applies to models of binding agreements, *even in a one-shot setting*. If some agents break off negotiations, surely the remaining agents are aware of this fact, so that repercussions of some sort are to expected? In contrast, this issue cannot arise in a one-shot model of nonbinding play: in effect, the actions of the complementary set of players are taken as "given".[6] With binding agreements, the very fact that a subgroup has moved off from the negotiating table allows the complementary group to change their behavior, and so it must be accounted for by the "deviating" subgroup, even in a "one-shot" theory.

Observe that the distinction between characteristic functions and the extended versions that incorporate externalities (discussed earlier) is important for item [2]. With characteristic functions, the complementary coalition may react to a coalitional deviation, but whether or not it does so is entirely irrelevant to the members of the deviating party : their worth does not depend on it. With externalities across players, however, the behavior of the complementary coalition becomes important.

2.5 Two Examples

I use two examples to illustrate some of the main points made so far. I will invoke these examples again at later stages in the book.

---

[4]Some of these words are placed in quotes because such refinements are not (necessarily) based on any sort of real-time dynamics; they are restrictions on the credibility of deviations.

[5]See, e.g., Aumann and Maschler (1964), Ray (1989), Dutta and Ray (1989, 1991), Dutta, Ray, Sengupta and Vohra (1989), Mas-Colell (1989) and Greenberg(1990).

[6]To be sure, similar considerations would crop up in noncooperative games when they are played over time.

2.5.1 Oligopoly. Several Cournot oligopolists produce output at a fixed unit cost, c, in a homogeneous market with a linear demand curve: $p = A - bx$. They are free to form coalitions among themselves, and this includes the option of forming the grand coalition of all players. Recall that by standard calculations, that the Nash profit accruing to a single firm in an m-player Cournot oligopoly is

$$\frac{(A - c)^2}{b(m + 1)^2} = \frac{D}{(m + 1)^2},$$

where $D \equiv (A - c)^2 / b$.

Indeed, this value is precisely the worth of a particular *coalition* if it is immersed in an environment with m coalitions altogether. Notice how different it is from a characteristic function. In such a function, the worth of a coalition would typically depend on the characteristics of that coalition. This example lies at the opposite extreme in that the worth of a coalition does not depend on the coalition itself, but on the number of all coalitions in society. This sort of "extended characteristic function" will be referred to as a *partition function*; worths will depend on the entire coalition structure.

Let us work briefly with this example without imposing any particular behavioral model just yet. Suppose that there are just three firms in all, deciding whether or not to form a cartel. If they do, they will earn monopoly profits, which from the expression above equals $D/4$. Obviously, in any proposed agreement between the three at least one of the firms must earn no more than $D/12$. What should this firm do?

Von Neumann and Morgenstern's characteristic function tells us that if this firm breaks off, it should anticipate whatever it is that the other firms can hold it down to. But this last number is zero, for it is certainly the case that the other two firms can flood the market and drive prices down to zero. So the characteristic function predicts that our firm should not object to *any* nonnegative return, however small. This is clearly absurd.

On the other hand, suppose that our firm anticipates that in the event of its defection, the other two firms will play a best response to the defector's subsequent actions. This implies that following the deviation, we are in a duopoly, where the deviant's return, using the general expression above is $D/9$. This exceeds $D/12$.

Does this mean a deviation from the three-player coalition is then justifiable? Not really: there are other considerations. Study the situation facing the two remaining firms once our deviant leaves. Their *total* return is $D/9$ as well, which means, of course, that one of them can be earning no more than $D/18$. If this firm were to leave and induce the standard three-person oligopoly, its return would be $D/16$. So faced with the irrevocable departure of one firm from the original agreement, the remaining firms will split up as well. But in this case, the original deviant gets $D/16$ too! So each member of the three-firm coalition should anticipate receiving $D/16$ as a result of such a deviation. It follows that the grand coalition in this example is a stable coalition structure (proposing the joint monopoly outcome with each firm getting at least $D/16$).

So this three-player example suggests that the grand coalition will indeed form, a result we would have obtained with characteristic functions as well. But the connection is entirely coincidental. Here is another three-agent example with markedly different implications.

2.5.2 Public Goods. Consider the provision of a public good by three symmetric agents. Suppose that each unit of a resource r (time, effort, money) contributed yields one unit of the public good, but generates a convex utility cost $(1/3)r^3$. As in the Cournot example above, let us construct a partition function for this game. A coalition of s players will contribute a per-capita amount $r(s)$ to maximize

$$sr(s) - \frac{1}{3}r(s)^3.$$

(In doing this — in writing down a single per-capita worth — I am implicitly assuming that utilities are freely transferable across the agents, say by linearly valued sidepayments of money.)

It is easy to see that $r(s) = \sqrt{s}$.

Now, the above expression does *not* represent the per-capita worth of the coalition. To evaluate the payoff of subcoalitions, however, we need to know production "elsewhere" and therefore the coalition structure "elsewhere". If production elsewhere is z, then the overall payoff to a coalition of size s is

$$s\left[z + sr(s) - \frac{1}{3}r(s)^3\right] = s\left[z + \frac{2}{3}s^{3/2}\right],$$

which can be solved more fully as soon as we know what the coalition structure elsewhere — and therefore z — is. Writing $\mathbf{v}(\pi)$ to be the (ordered) vector of coalitional worths in the coalition structure π, it is easy to see that

$$\mathbf{v}(\{123\}) = \{6\sqrt{3}\}$$

$$\mathbf{v}(\{1\}, \{2\}, \{3\}) = \left\{2\frac{2}{3}, 2\frac{2}{3}, 2\frac{2}{3}\right\}$$

$$\mathbf{v}(\{i\}, \{jk\}) = \left\{2\sqrt{2} + \frac{2}{3}, 2\left[1 + \frac{2}{3}\sqrt{8}\right]\right\}.$$

The exact numbers in this array are unimportant. Just as in the case of the Cournot example, we shall be providing a more general analysis later in the book. There are only two features to be noted here. First, the per-capita worth of the grand coalition is $2\sqrt{3}$, which is smaller than the payoff to i in the coalition structure $\{\{i\}, \{jk\}\}$, which is $2\sqrt{2} + \frac{2}{3}$. This is in keeping with the Cournot example. But the second feature is different: the per-capita payoff to j and k in the coalition structure $\{i\}, \{jk\}$, which is $[1 + \frac{2}{3}\sqrt{8}]$, exceeds their corresponding payoff in the coalition structure of singletons, which is $2\frac{2}{3}$. In contrast to the Cournot example, if one agent commits to (irreversibly) exit the negotiations, it is in the interest of the remaining two players to stay together.

So in this case, and in contrast to the first example, it is difficult to avoid an inefficient outcome. The grand coalition is not capable of arranging a payoff allocation that will cater to every conceivable single-person threat to "boycott" the negotiations. And that single-person threat has credibility: faced with the imminent and irrevocable departure of one agent, the remaining agents will find it in their best interests to cling together.

The examples illustrate well the use of partition functions that capture externalities across players. The payoff to a single "deviant" from the grand coalition is not defined in isolation: it is fully contextual and depends on what the other players do. A theory that encompasses such externalities will automatically be able to handle a variety of real-world applications, something that the traditional characteristic function cannot do.

Closely related to this point is the role played by farsightedness. A player's payoff is linked to the ambient coalition structure, so some degree of understanding of the consequences of an action is necessary for a satisfactory theory (at least in a model with rational players). Recall Aumann and Myerson: "Where will it all lead? Is the *end result* good or bad for me?" As we've seen, farsighted players arrive at very different conclusions in the two examples, while myopic players wouldn't.

2.6 Negotiations: One-Time or Continuing?

The two examples suggest another fundamental consideration. Take, for instance, the public goods example. When one agent moves away, he expects that the other two agents will stick together because what they get together exceeds the payoffs to them of remaining apart. Indeed, this is what encourages the initial agent to free-ride in the first place.

But matters are not that simple if the first agent's commitment to leave is not entirely irreversible. Now agents 2 and 3 can get him back, and this gives rise to all sorts of additional considerations. For instance, agents 2 and 3 may now make a joint proposal to agent 1 asking him to "return" to the grand coalition in return for an even larger payoff: if utilities are transferable this is a clear possibility. Or in an even more farsighted stroke, agents 2 and 3 may *also* commit to break up, in the hope that in future periods this will force agent 1 back into the fold on a more symmetric basis.

Our model of negotiations permits coalitions to form freely, and divide their payoffs freely (if utility is transferable). But it is far more circumspect on the question of *further* negotiations. One possible view is that a commitment once made is irreversible. Another view is that no commitment is so strong that it cannot be reversed.

What is the "right model", then? The answer is that there is no right model. It depends on something that we typically blackbox in our abstract game-theoretic exercises, but now needs to be brought out into the open: the *technology of commitment.*

There are three different kinds of working models we can write down about commitments. First, one might assume irreversible commitments. This will be true of situations in which commitment must be made by the use of concrete actions (not legal devices) that are prohibitively costly to reverse. An obvious example is one in which groups form to engage in violent conflict: the call to hostilities may be pretty much irreversible, at least in the short run.

Or consider the public goods example of the previous section. Say that the good in question is pollution control. Such control might require the setting up of environmentally friendly factories that must be built from scratch. A firm (or region, or country) that does *not* take this route is committing to a increased level of pollution (or equivalently, a lower production of the public good) that may be too costly to reverse. For instance, it may be setting up factories that are not built environment-friendly, and moving to a greater level of control will require the tearing-down of these fixed investments.

Second, a commitment may be costly but reversible (at some additional cost). Continuing in the vein of the previous example, this might occur if pollution-control devices are, by and large, modular, so that they can be tagged on to existing installations at moderate cost. In this situation renegotiation is more likely to occur, but nevertheless the inefficient situation described in the example must first obtain, *otherwise the beneficiaries of renegotiation will have no power to extract the surplus.*

Finally, it may be that a commitment, once made, is costless to reverse. This may sound a bit paradoxical (and perhaps it is), but let us pursue this line of thought a bit further. In the context of our public goods example, it may be that the first player has access to a legal commitment device in which it makes the following declaration: that it commits to form no binding relationship with the other players, *unless* the other players are signatories to an agreement that gives it more than its stand-alone payoff. The qualification may then be used to "reverse the commitment", which was really a conditional commitment all along.

Alas, I need to irreversibly commit what goes into this book, so I will hedge a bit and study the two extreme cases. In the chapters that follow, we will first study the case in which commitments once made cannot be altered. Then we switch gears and introduce a dynamic model of binding agreements, in which earlier commitments can be reversed. I will do this twice, once for the bargaining approach in Part 2 and once for the blocking approach in Part 3.

CHAPTER 3

Coalitions, Cooperation and Noncooperation

3.1 Definitions and Notation

The background for all we do is a game Γ in strategic form. N is a set of players. Player i has action set A_i. $\mathbf{A}$ denotes the product of all action sets. Player i has a payoff function u_i defined on $\mathbf{A}$.

Our approach is fundamentally different from that of noncooperative game theory. We presume that binding agreements can be written and implemented. The problem is: which agreements will be written, and what is the resulting structure of coalitions that will emerge?

A *coalition* is just a nonempty subset of N. We interpret a coalition to mean a set of players who are willing signatories to a binding agreement. We will have occasion later to discuss the implications of this interpretation, some (but not all) of them semantic, but let's not muddy the waters just yet.

The restriction of the product set $\mathbf{A}$ to coalition S will be denoted by $\mathbf{A}_S$. Similarly, for $\mathbf{a} \in \mathbf{A}$ and coalition S, denote $\{u_i(\mathbf{a})\}_{i \in S}$ by $\mathbf{u}_S(\mathbf{a})$. The complement of a coalition S will be denoted by $-S$.

A partition of N into coalitions, denoted by π, will be called a *coalition structure*.

3.2 Coalitional Equilibrium

3.2.1 Definition. To figure out what agreements will be written, we need to study the flip side: what happens if an agreement is *not* forthcoming? The implicit idea, then, is one of "two stages": each player must forecast the interactive consequences of every conceivable coalition structure. Coalition formation occurs "at an earlier stage" with this "second-stage consequence" firmly in mind.

Indeed, it is to summarize this "second stage" that characteristic functions typically make an appearance, by invoking criteria to describe what coalitions (including individuals) can guarantee themselves.

In contrast, we describe noncooperative interaction *across* coalitions entirely in the spirit of Nash. Suppose that π is a coalition structure. An action vector $\mathbf{a}$ is a *coalitional equilibrium* (relative to π) if for no coalition $S \in \pi$ is there an action vector $\mathbf{a}'_S \in \mathbf{A}_S$ with $\mathbf{u}_S(\mathbf{a}'_S, \mathbf{a}_{-S}) \gg \mathbf{u}_S(\mathbf{a})$.

A coalitional equilibrium is a natural extension of Nash equilibrium: simply treat different coalitions as agents with vector-valued payoffs. The idea is simple: once a coalition structure is in place, each coalition in that structure has agreed to write binding agreements among its members, while across coalitions there is no such arrangement.

For singleton coalition structures a coalitional equilibrium is just a Nash equilibrium, and for the grand coalition it is simply the Pareto frontier of the game.

A coalitional equilibrium has little to do with noncooperative coalitional refinements of Nash equilibrium, such as coalition-proofness, or with hybrid concepts such as strong equilibrium. It is simply presumed that coalitions within a given structure can write any binding agreement they please (just which one they *do* write will be the main focus of our study later), and that across coalitions there is noncooperative play. In particular, a coalitional equilibrium is *not* a refinement of Nash equilibrium for the underlying strategic form game Γ.

3.2.2 Existence. The existence properties of coalitional equilibria are entirely unremarkable. Ray and Vohra (1997) establish the following result.

PROPOSITION 3.1. *Suppose for all i, A_i is nonempty, compact and convex and u_i is continuous and quasi-concave. Then for every coalition structure, a coalitional equilibrium exists.*

Proof. Fix some coalition structure π. For each $S \in \pi$, define the "better-than set" relative to $\mathbf{a} \in \mathbf{A}$ as follows:

$$P_S(\mathbf{a}) \equiv \{\mathbf{a}'_S \in \mathbf{A}_S \mid \mathbf{u}_S(\mathbf{a}'_S, \mathbf{a}_{-S}) \gg \mathbf{u}_S(\mathbf{a})\}.$$

The quasi-concavity of payoffs assures us that $P_S(\mathbf{a})$ is convex for all S and x. While the underlying ordering is not complete, it is easy to see that the graph of P_S is open. Invoking the existence result of Shafer and Sonnenschein (1975), there exists $\bar{\mathbf{a}} \in \mathbf{A}$ satisfying the required properties of a coalitional equilibrium. □

The notion of a coalitional equilibrium has precursors in the work of Ichiishi (1981), Myerson (1991, p .424) and Zhao (1992).[1] Haeringer (2004) addresses questions of existence when the quasi-concavity of payoffs is not assumed.

3.2.3 Interpretation. Our implicit insistence that coalition formation will be followed by the play of some coalitional equilibrium is restrictive, especially in dynamic settings. It is entirely possible that conditional on the formation of a particular coalition structure, the coalitions in it might follow history-dependent actions, just as in the theory of repeated games. Because we presume that all agreements that need to be supported can in fact be supported by binding agreements, we neglect this aspect. Accordingly, in all that we do, the formation of a particular coalition structure will be followed by the play of some equilibrium relative to that structure. To be sure, the structure itself can change over time, and then so will the associated action profile.

3.2.4 Essential Uniqueness. Coalitional equilibria need not be unique. After all, Nash equilibria, which are just coalitional equilibria for singleton coalition structures, are generally not unique.

---

[1]Myerson (1991, Section 9.2) contains a succinct description and comparison of different approaches to deriving characteristic functions, including the "rational threats" approach due to Harsanyi (1963).

Coalitional equilibria must therefore be prey to exactly the same sort of multiplicity.

That said, there is a plethora of applied situations in which a unique Nash equilibrium does exist. Because the theories we develop need to be subject to careful scrutiny on several grounds, we will refrain from compounding the difficulties by entertaining this sort of multiplicity, and the attendant conceptual complexities involved in equilibrium prediction in this case.

However, there is a second source of nonuniqueness which is particular to coalitions. If payoffs are transferable to some degree across players, there can be many payoff vectors attached to a particular coalition structure, simply because each coalition can divide up its aggregate payoff in different ways. It would be extremely restrictive to rule out *this* source of multiplicity. In doing so, we would be closing off all avenues for transfers across players, which is problematic to say the least.

Fortunately, the first sort of multiplicity is easily distinguishable from the second. Let $\mathbf{U}(\pi) \subseteq \mathbb{R}^N$ denote the set of coalitional equilibrium payoffs associated with any coalition structure. Say that coalitional equilibrium is *essentially unique* if there exist sets $\mathbf{U}(S, \pi) \subseteq \mathbb{R}^S$ for every coalition $S \in \pi$ such that

$$(3.1) \qquad \mathbf{U}(\pi) = \bigtimes_{S \in \pi} \mathbf{U}(S, \pi).$$

The product structure of "essential uniqueness" allows us to distinguish between the two kinds of multiplicity. Nash-equilibrium-like multiplicity will typically not display a product structure. On the other hand, the multiplicity that simply arises from alternative intra-coalitional "divisions" of payoff will satisfy the essential uniqueness property.

For instance, all characteristic functions trivially display essential uniqueness. But as we shall see, so do a host of other economic models with genuine externalities.

In the rest of this monograph, we place emphasis on games of coalition formation whose coalitional equilibria satisfy the essential uniqueness property for every coalition structure.

3.3 Partition Functions

3.3.1 Definition. Coalitional equilibria that satisfy the property of essential uniqueness give rise to what we will call *partition functions*, descriptions of the payoff possibilities accruing to each coalition *within the context of a particular coalition structure*. A partition function is generated by precisely the sets described in (3.1). For every coalition structure π and every coalition $S \in \pi$, it assigns a set of payoff vectors $\mathbf{U}(S, \pi)$, describing the equilibrium opportunities available to that coalition when embedded in the structure π.

A partition function becomes a characteristic function in those cases in which the dependence of $\mathbf{U}$ on π can be removed. Now there are genuinely no interactions across coalitions, so that we can define coalitional worth "in isolation," without needing to be aware of the surrounding context.

If the underlying strategic game has interpersonally comparable utilities, and if side payments can be made across subsets of players, the set of all payoffs to a coalition could then be identified with a single number, its *worth*. Formally, for every structure π and coalition $S \in \pi$, we would have a function $v(S, \pi)$ such that

$$\mathbf{U}(S, \pi) = \left\{ \mathbf{u} \in \mathbb{R} \Big| \sum_{i \in S} u_i \leq v(S, \pi) \right\}.$$

Such a TU partition function is precisely analogous to a TU characteristic function, except that the richer dependence on the underlying structure π continues to be maintained.

In this book, we often begin with the partition function as a primitive. It is not really necessary for us to suppose that there is, in fact, some underlying strategic form game which can be "reduced" to a partition function in the way described above. Indeed, to the extent that such underlying games impose structure on partition functions, we may be tying our hands too tightly.

As an example: a partition function as derived from the strategic form *must* satisfy "grand coalition superadditivity". Consider a payoff vector $\mathbf{u}$ "pieced together" from a particular partition π, so that $\mathbf{u}_S \in \mathbf{U}(S, \pi)$ for every $S \in \pi$. Then it must be the case that there exists $\mathbf{u}' \in \mathbf{U}(N, \{N\})$ such that $\mathbf{u}' \geq \mathbf{u}$. In short, the grand coalition can do everything a partition can.

Is that a good assumption? Very often it is, but very often it isn't. See Section 3.4 and Chapter 10 for more.

3.3.2 Examples. In this section, I list several examples of characteristic and partition functions. The examples are often impressionistic and not presented in a rigorous format. However, in the chapters that follow, I return to some of these examples and study them in more detail.

1. *Local Public Goods.* There are n people and a large number of locations in which they can choose to live. Person i gets utility $f(c) + h(g)$, where c is money and g is a local public good, produced from individual contributions in person i's locality. That is, $g = g(T)$, where $T = \sum_{k \in S}(w_k - c_k)$, and w_k is the money endowment of player k.

The action set A_i is constructed as follows. Each person chooses three things: one of a large finite set of locations, a nonnegative vector of money contributions to each location, and a nonnegative vector of money transfers to every other player. Generate consumption c_i by subtracting all these transfers and contributions, but add any transfers that i may receive from others. Generate g by adding all contributions to the local public good in the locality chosen by i. Let $u_i = f(c_i) + h(g)$.

In a coalitional equilibrium, it is obvious that there is no interaction across coalitions. Each member in a coalition selects the same location, and no transfers are made to any other person or locality. We therefore have a characteristic function $\mathbf{U}(S)$. Solve for it — or rather its Pareto frontier — by addressing the following problem: For any i and for arbitrary numbers y_j, $j \in S$, $j \neq i$,

$$\max f(c_i) + h(g)$$

subject to $f(c_j) + h(g) \geq y_j$ for all $j \neq i$, and the feasibility constraint $g = g(T)$, where $T = \sum_{k \in S}(w_k - c_k)$.

If $f(c)) = c$, we are in the quasi-linear case and this yields a TU characteristic function. The worth of each coalition is obtained simply by maximizing

$$\sum_{i \in S} c_i + sh(g)$$

(where s is the size of S), subject to the feasibility constraints described above.

2. *Global Public Goods.* Consider the same example as before, except now there is no choice of localities. This seemingly simpler problem is actually more complicated, for now there are externalities across coalitions.

Once again, in a coalitional equilibrium, no transfers are made across coalitions. For ease in exposition, let us restrict ourselves to the TU case. To calculate the best response of coalition S, simply maximize, for given aggregate contributions $\overline{T}$ from the complement $-S$,

$$\sum_{i \in S} c_i + sh(g)$$

subject to $g = g(T)$ and $T = \sum_{k \in S}(w_k - c_k) + \overline{T}$.

It is not hard to verify that these best responses yield coalitional equilibrium payoffs that are essentially unique. We therefore have a well-defined TU partition function. A similar analysis is easily conducted for the NTU case.

For more on coalition formation in public goods economies, see Ray and Vohra (2001) and Section 6.2.

3. *Winning Coalitions.* A special subcollection of coalitions can win an election, whereupon they get one unit of surplus. So $v(S) = 1$ for every coalition, and $v(S) = 0$ otherwise.

The subgroup may be defined by a supermajority: for instance, $v(S) = 1$ if and only if $s \geq m$, where $m > n/2$ is the supermajority threshold.

Winning coalitions also make an appearance in n-person bargaining games that are settled by majority rule (see, e.g., Baron and Ferejohn (1989)). In Chapter 7, we use this idea to embed majority-rule bargaining games into coalitional bargaining situations with unanimity (see also the discussion in Section 14.1 of Chapter 14).

4. *Conflict.* There are several individuals. Each coalition in a coalition structure of individuals expends resources to obtain a reward (perhaps the pleasures of political office as in the previous example). Resources may be spent to lobby, finance campaigns, or engage in cross-coalitional conflict, depending on the particular application. Suppose that the probability p_S that coalition S wins

depends on the relative share of resources r expended by it:

$$(3.2) \qquad\qquad p_S = \frac{r_S}{r_S + r_{-S}}.$$

The per-capita value of the win will generally depend on the characteristics of the coalition (for instance, coalitional size); write this value as w_S. In the TU case, the coalition then chooses resource contributions from its members to maximize

$$sp_S w_S - \sum_{i \in S} c(r_i),$$

where c is the individual cost function of contributions, $r_S = \sum_{i \in S} r_i$, and r_{-S} is taken as given.

It is easy to check that essential uniqueness is satisfied, so that we have a well-defined TU partition function. The NTU case is easy to handle using a similar construction.

Esteban and Ray (1999) study a more general version of this conflict model. For models of coalition formation based on this framework, see Esteban and Sákovics (2004) and Bloch, Soubeyran and Sánchez-Pagés (2006).

5. *Production Teams.* Each agent has "ability" α_i. When a group S of agents gets together, they can produce an output $= f_S(\alpha_S)$ (where f_s is a family of functions indexed by s).

If the output is sold at a fixed price that's independent of the set of agents under consideration (or just plain consumed), this generates a characteristic function. It will be TU if the proceeds can be split in any way and agent utilities are linear in those proceeds. If, in contrast, there is a fixed sharing rule (such as equal division), the resulting characteristic function will be NTU.

6. *Oligopolies: Production Teams With Interaction.* A given number (n) of Cournot oligopolists produce output at a fixed unit cost, c. The product market is homogeneous with a linear demand curve: $p = A - bx$. Recall that by standard calculations, the payoff to a single firm in an m-firm Cournot oligopoly is

$$\frac{(A - c)^2}{b(m + 1)^2}.$$

If each "firm" is actually a cartel of firms, the formula is no different as long as each cartel attempts to maximize its total profits (and then freely allocated these profits among its members). once again, essential uniqueness is satisfied, and

$$v(S, \pi) = \frac{(A - c)^2}{b \, [m(\pi) + 1]^2},$$

where $m(\pi)$ is the number of cartels in the coalition structure π. This partition function is particularly interesting, in that it does not depend on S at all, but only on the ambient coalition. In this sense the Cournot oligopoly exhibits properties that are as far removed from traditional characteristic functions as possible.

For more on oligopolies, see Bloch (1996), Ray and Vohra (1997, 1999) and Section 6.1.

7. *Political Coalitions.* There are n politicians, each representing an equal measure of (a continuum of) voters. The set of feasible policies is some set Q. Voters in group i share the same preferences as their political representative: $u(q, i)$ on Q.

For every coalition structure — a partition of all politicians into "parties" — the parties contest a general election. Each party can announce a platform $q \in Q$, or some "null platform" $\emptyset$ which may be interpreted as not running. Voters vote sincerely across all platforms, and the platform with the highest vote wins (ties are broken randomly).

The Nash equilibrium of this game generates a partition function. For more on this game, see Levy (2004) and Section 12.4 of Chapter 12.

8. *Customs Unions.* There are n countries, each specialized in the production of a single good. There is a continuum of consumers equally dispersed through these countries. They all have identical preferences. Impose the restriction that no country or coalition can interfere with the workings of the price system except via the use of import tariffs. Then for each coalition structure — a partition of the world into *customs unions* — there is a coalitional equilibrium, in which each customs union chooses an optimal tariff on goods imported into it.

In particular, the grand coalition of all countries will stand for the free-trade equilibrium: a tariff of zero will be imposed if lump-sum transfers are permitted within unions.

A specification of trade equilibrium for every coalition structure generates a partition function for the customs union problem.

For different models of customs union (from an explicit coalition-formation context), see Krugman (1993), Krishna (1998), Ray (1998, Chapter 18), and Aghion, Antras and Helpman (2004).

3.4 Extensions

There are two restrictions (surely others, but two I particularly care about) that are implicit in the development of this chapter. I want to briefly address them right away. I will be returning to both these restrictions at later points in this book.

First, I've defined a coalition structure to be a *partition*. This means that each individual writes agreements with at most one group of players: the coalition to which she belongs. If she does not write any agreements, she forms a "singleton coalition", which is certainly allowed for by the partitional restriction. What is *not* allowed for is the possibility that coalitions may overlap, so that some individuals write two (or more) different agreements with different groups.

A good example of this scenario is the writing of regional trade agreements. A country can be party to more than one trade agreement with different subsets of partners. (This would not be the case for a customs union.) One would then need to define the notion of a *cover*: a possibly overlapping collection of coalitions that exhaust the player set.

My monograph will not deal with these issues, though I return to them in Chapter 14 (see Section 14.4). I conjecture, however, that the techniques developed here will be extremely useful in addressing such problems at some point down the road.

In fortunate contrast, the second restriction will not hinder what follows. First the restriction: it lies in the implicit assumption that *everybody*, in principle, can enter into an full agreement with everybody else. There are many interesting and important problems in which this is not the case. A group of domestic firms might consider the formation of a cartel, but such agreements may be ruled out between domestic and foreign firms exporting to the same market. Two or more ethnic groups may form an alliance with one another, but never with a common enemy with whom they share a long history of animosity. Company shareholders may always wish to spin off some divisions because healthy competition among those divisions cannot be maintained under the umbrella of one company name. Factors as varied as antitrust laws, ideology, ancient hatreds, geography or the competitive spirit may conspire to rule out certain coalitions (and not others).

In contrast, we derive partition functions from strategic games under the presumption that *once* a coalition has formed, "cooperation" within that coalition is unrestricted. This has the important implication that a partition function must be grand coalition superadditive (see Section 3.3 for the definition). In none of the examples of the previous paragraph is this necessarily the case. If cooperation between coalitions is restricted or simply unavailable, one must must derive partition functions from strategic forms under these additional restrictions.

While it isn't hard to extend the argument to accommodate such cases, we won't need to do that in this book because most of it treats the partition function as a primitive. What is more, in some of the important examples (see Chapter 10 in particular) we deliberately invoke the possibility that the partition function may fail to be superadditive. Section 10.3 in Chapter 10 will continue this discussion.

3.5 Summary

In this chapter, we've introduced the concept of coalitional equilibrium and used that concept to derive partition functions. In what follows, partition functions will present our basic building blocks, but it is important to remember that they are, in turn, reduced-form objects.

Partition functions embody several important restrictions. First, the use of partition functions implicitly presumes that conditional on the formation of a particular coalition structure, a coalitional equilibrium will indeed be played. This means that we employ a neat separation between the process of coalition formation and the ability to *also commit to a course of action in that process*. If there is a Stackelberg structure (or more generally, some form of sequentiality) to the game, that is presumed to be already embedded in the description of the underlying strategic form. In short, whether or not you get to form a coalition before someone else, you are presumed to choose a binding agreement for your coalition that is a "best response" to the actions of other coalitions in the ambient coalition structure. No further sequential commitment is permitted on that front.

Second, partition functions presume that the essential uniqueness condition is met. The exact reasons for this restriction, while not as important as the one discussed earlier, will become clearer later.

Partition functions also assume that a coalition structure invariably takes the form of groups that do not overlap, so that no player enters into binding agreements with a variety of groups. This is a restriction of our analysis but I expect the methods in the book to extend to this more general case (Section 14.4 of Chapter 14 sketches a possible extension).

Finally, the very process of deriving partition functions from strategic form games may on occasion be restrictive (see the discussion in Section 3.4). While we do carry out such a derivation in this chapter, we neither insist on nor require this interpretation in the sequel. Partition functions are derived objects all right, but they can also serve as good primitives.

Part 2

A Bargaining Approach To Coalition Formation

CHAPTER 4

Irreversible Agreements

4.1 Introduction

In this chapter, we introduce a model of binding agreements and coalition formation with two important features. First, all negotiations are expressed formally as a *bargaining game*, which we base on Ståhl (1977), Rubinstein (1982) and Chatterjee, Dutta, Ray and Sengupta (1993).[1] We've already discussed the pros and cons of entertaining such a well-defined extensive form; the objective now is not further soul-searching but an examination of where this assumption might take us. Second, we assume that all agreements to form a coalition are "fully" binding, in the sense that they are irreversible. A coalition once formed cannot disintegrate or be subsequently absorbed into a larger group.

Both these assumptions will be dropped in due course (the second first, the first later) but it is important to appreciate that the assumptions are not necessarily to be viewed as restrictions. Sometimes, a well-defined negotiation process may be a better approximation to reality than the more eclectic blocking models, and sometimes an irreversible commitment may be a far better description of the situation at hand than a model which allows all prior commitments to be reversed.

Throughout, we regard the partition function as a primitive, with the idea that underlying this function is a game in strategic form (recall Chapter 3). We define on the partition function a noncooperative

---

[1]For related literature on bargaining, see Binmore (1985), Moldovanu (1992), Okada (1996), Perry and Reny (1993), Selten (1981) and Winter (1993).

bargaining game. Proposers offer to form coalitions, and to divide coalitional worth in particular ways. Responders agree or disagree. Coalitions form through the course of this bargaining process.

The analysis in this chapter and the two chapters to follow draws heavily on Chatterjee, Dutta, Ray and Sengupta (1993) and Ray and Vohra (1999), though we extend these papers along several dimensions.

4.2 A Model

4.2.1 Preliminaries. $N = \{1 \dots n\}$ is the set of *players*. A *coalition structure of N* is a partition π of N. A *partition function* $\mathbf{U}$ assigns to each coalition S in a coalition structure π a compact set of nonnegative payoff vectors $\mathbf{U}(S, \pi)$.[2]

We impose the mild restriction that $\mathbf{U}$ is *comprehensive*. Essentially, this asks for *some* transferability of payoffs along the efficient frontier of worths, though to arbitrarily small degree. Formally, we suppose that for every π and $S \in \pi$, whenever $\mathbf{u} \in \mathbf{U}(S, \pi)$ has $u_i > 0$ for some i, then there is $\mathbf{u}' \in \mathbf{U}(S, \pi)$ with $u'_j > u_j$ for all $j \neq i$. It is easy to see that the frontier of $\mathbf{U}(S, \pi)$ is now "downward sloping" everywhere.

4.2.2 Proposals and Responses. Agents make proposals to coalitions and respond to proposals made to coalitions to which they belong. Imagine that some coalitions have already formed. To each "remaining" set of "active" players is assigned some probability distribution over initial proposers. Likewise, to each coalition of active agents *to* which a proposal has been made, there is a given order of respondents (excluding the proposer of course).

Loosely speaking, a proposal is the division of the worth of a coalition among its members. But given a partition function, a worth is not well-defined until a coalition structure has formed in its entirety. Therefore a proposal must consist of a set of *conditional statements* that describe a proposed division of coalitional worth for every contingency; i.e., for every conceivable coalition structure that finally forms. More precisely, a proposal is a pair $(S, \mathbf{u})$, where $\mathbf{u}$ is a collection of allocations $\{\mathbf{u}(\pi)\}$, one for each partition π that contains

---

[2]The nonnegativity is just a normalization, though the payoff from eternal disagreement will have to be suitably restricted; see below.

S, feasible in the sense that for every coalition S in π,

$$\mathbf{u}_S(\pi) \in \mathbf{U}(S, \pi).$$

We assume that a proposer must include herself in the proposed coalition. Under some mild conditions, this requirement is unimportant, and we will discuss it later.

Once a proposal is made to a coalition, attention shifts to the respondents in that coalition. A response is simply acceptance or rejection of the going proposal. If all respondents unanimously accept, the newly-formed coalition exits, and the process shifts to the set of still-active players remaining in the game.

The unanimity requirement appears restrictive in some situations in which a majority or (nonunanimous) supermajority is enough to implement a proposal. To see why this restriction is illusory, see Section 4.5, though Section 14.1 of Chapter 14 adds some caveats.

The rejection of a proposal creates a bargaining friction: payoffs are delayed by the passage of some time, which is discounted by everybody using a discount factor δ. What happens next typically varies from model to model. We adopt a fairly general specification for now, but it will be tightened later and different variants discussed in more detail. We presume that the rejector can choose to leave the bargaining table, effectively forming a one-person coalition. If she does not exit, she gets to be the next proposer with some probability $\rho(s)$ that may depend on the number of active agents s. With probability $1 - \rho(s)$, some other active agent is equiprobably chosen to be the new proposer. Later, I will need to place more restrictions on the protocol, but for now this will do.[3]

If and when all agreements are concluded, a coalition structure forms. Each coalition in this structure is now required to allocate its worth among its members as dictated by the proposals to which they were signatories. If bargaining continues forever, it is assumed that all the "stalled" players receive some strictly negative

---

[3] I could have actually written down an even more general proposer protocol for which all the results presented here would be valid. For instance, I could allow the proposer probabilities to depend quite generally on the history of the game so far. But nothing much would be gained thereby except an increase in notation, so I avoid this.

payoff.[4] (Nothing of substance is affected by allowing already-formed coalitions in such cases to receive any payoffs we please.)

4.2.3 Strategies and Equilibrium.

A *strategy* for a player requires her to make a proposal whenever it is her turn to do so, where the choice of proposal could depend on events that have already unfolded. It also requires her to accept or reject proposals at every stage in which she is required to respond. A *perfect equilibrium* is a profile of strategies such that there is no history at which a player benefits by deviating from her prescribed strategy.

It is well known from the theory of repeated games that perfect equilibria can generate a huge multiplicity of outcomes. Bargaining games — while potentially infinite — are *not* repeated games. Yet, as Section 4.4.3 makes clear, the use of history-dependent strategies also generates a multiplicity of outcomes. Later, when we study coalition formation in real time, we shall have more to say about multiplicity and history-dependence. But provisionally, and for the purpose of this chapter, we accept such multiplicity and retreat to the use of *stationary Markovian strategies*.

Stationary Markovian strategies depend on a small set of state variables, and do so in a way that's insensitive to the passage of calendar time. The current proposal or response (while permitted to be probabilistic in nature) is not permitted to depend on "past history". Of course, it must be allowed to depend on the current set of active players, on the coalition structure that has already formed, and — in the case of a response — on the going proposal. We will also permit proposers to condition their new proposals on the identity of the last rejector (in the current round of negotiations), and for respondents to condition their responses on the identity of the proposer.

A *stationary Markovian equilibrium* is then a collection of stationary Markovian strategies which forms a perfect equilibrium. This completes the description of the model.

---

[4]The word "negative" is relative to the normalization that all coalitional payoffs are nonnegative; see above, footnote 2.

4.3 Equilibrium

Our notion of equilibrium allows for mixed (behavior) strategies. In fact, it does so in three ways: (a) the proposer may randomly choose a coalition, (b) given the choice of a coalition, the proposer may randomly choose offers, and (c) respondents may mix over accepting and rejecting a proposal.

But it turns out that a stationary equilibrium exists with a minimal need to randomize, as described in the proposition below.

PROPOSITION 4.1. *If* **U** *is compact and comprehensive, there exists a stationary Markovian equilibrium in which bargaining ends in finite time.*

The proof of this proposition is technical and therefore relegated to an appendix to this chapter.[5]

4.4 Rubinstein–Ståhl Bargaining

For later reference, I briefly describe a well-known noncooperative bargaining model based on Rubinstein (1982) and Ståhl (1977). In this model several persons divide a cake of size 1; there are no subcoalitions of any value, and there are no externalities.

This model can be embedded very easily into our framework. Consider the following simple TU characteristic function: $N = \{1 \dots n\}$, $v(N) = 1$, while $v(S) = 0$ for subcoalitions S. The protocols we consider are in line with the class discussed above: an initial proposer is chosen with some probability, the latest rejector gets to propose with probability $\mu > 0$, and everyone else with probability $(1-\mu)/(n-1)$ each. (Rubinstein studies alternating-offers bargaining when $n = 2$, so that $\mu = 1$ in that case.)

4.4.1 Two Persons. Suppose that $n = 2$, so that two persons are bargaining. A remarkable property of this two-person model is that subgame perfection fully pins down equilibrium payoffs.

---

[5]The proof, which relies on an inductive fixed point argument, may be of interest. At every subgame, a suitable fixed point (in payoff space) is constructed, and this fixed point replaces the relevant portion of the game, as we inductively move to an earlier subgame. To complete the fixed point argument for the earlier subgame, we need an additional continuity argument for the recursively constructed fixed points, which is where the possibility of mixing makes an appearance.

The proposition that follows is well-known from Rubinstein (1982), though I've generalized it to a broader class of protocols.[6]

PROPOSITION 4.2. *There is a unique subgame perfect equilibrium payoff vector in the two-person bargaining model.*

Proof. Existence will be shown below; assume it for now and prove uniqueness. Let M and m be the supremum and infimum equilibrium payoff to a *responder*, conditional on her rejecting the current offer but before the proposer has been decided.[7] Then it is obvious that a *proposer* can always assure herself an infimum of at least $1 - M$, so that

$$m \geq \delta[\mu(1 - M) + (1 - \mu)m],$$

or

(4.1)
$$m \geq \frac{\delta\mu(1 - M)}{1 - \delta(1 - \mu)}.$$

At the same time, no proposer can obtain more than $1 - m$, so it is *also* true that

$$M \leq \delta[\mu(1 - m) + (1 - \mu)M].$$

or

(4.2)
$$M \leq \frac{\delta\mu(1 - m)}{1 - \delta(1 - \mu)}.$$

Combining (4.1) and (4.2), it is easy to see that

$$m \geq \frac{\delta\mu\left(1 - \frac{\delta\mu(1-m)}{1-\delta(1-\mu)}\right)}{1 - \delta(1 - \mu)},$$

and simplifying this yields the inequality

(4.3)
$$m \geq \frac{\delta\mu}{1 - \delta(1 - 2\mu)}.$$

Following an analogous line of reasoning,

(4.4)
$$M \leq \frac{\delta\mu}{1 - \delta(1 - 2\mu)}.$$

---

[6]On the other hand, I assume a common discount factor. Dropping this restriction makes no difference to the argument.

[7]When discount factors are not the same, these values vary across the players but the proof follows exactly the same lines.

and together (4.3) and (4.4) show that

$$(4.5) \qquad M = m = \frac{\delta\mu}{1 - \delta(1 - 2\mu)} \equiv m^*,$$

which establishes uniqueness.

Existence can now be shown by construction. Have each player accept an offer if it yields her at least m^* (defined in (4.5), and always make the proposal $(1-m^*, m^*)$ when it is her turn to propose. It is easy to verify that this strategy profile constitutes a perfect equilibrium.

□

This proposition and its accompanying proof reveal that the equilibrium involves immediate agreement, with the proposer and the responder receiving

$$\frac{1 - \delta(1 - \mu)}{1 - \delta(1 - 2\mu)} \quad \text{and} \quad \frac{\delta\mu}{1 - \delta(1 - 2\mu)}$$

respectively. It is worth noting that no matter how small μ is, as long as it is strictly positive, the division of the cake must converge to an equal split as "bargaining frictions" vanish; i.e., as δ converges to 1. It is true that the first individual to propose may acquire a lot of power, especially if μ is small, but the value of that added power becomes negligible provided both players are extremely patient.[8]

4.4.2 More Than Two Persons. First consider stationary, symmetric strategy profiles. In such profiles, each person uses a common *response threshold* m, to be interpreted as the minimum amount for

---

[8]The fact that a unique and — for the two-person symmetric case at least — entirely sensible limit is selected (as bargaining frictions vanish) is comforting, and pushes us to ask what the corresponding limit would look like in general, something that we shall explore in more detail. In writing these lines, I am fully aware that with a slightly different specification of frictions — different discount factors, for instance — one might pick out a variety of other limit points. One might even interpret this feature as a failure of the theory, but that would be too literal an interpretation of the model. In my view, the homogeneous discount factor specification picks out a perfectly reasonable limit for the symmetric Rubinstein–Ståhl model — equal division — as bargaining frictions vanish. That makes it a natural theoretical benchmark. It yields the commonsensical prediction for a situation that we intuitively understand and allows us to explore more complex and unintuitive cases, such as general characteristic functions.

which she will accept a proposal, provided that all responders *after* her in the response order are planning to accept that proposal.[9]

In equilibrium, m must be built from an expectation about payoffs conditional on rejection; these would be a probabilistic combination of proposer payoffs $(1 - (n - 1)m)$ and responder payoffs (m). Therefore, m must solve the following equation:

$$m = \delta\{\mu[1 - (n - 1)m] + (1 - \mu)m\},$$

which tells us that

(4.6)
$$m = \frac{\delta\mu}{(1 - \delta) + \delta\mu n}.$$

This solution extends the two-person case and once again, convergence occurs to equal division as bargaining frictions disappear, provided that $\mu > 0$.

I've been a bit cavalier about deriving this stationary equilibrium. In particular, I took the shortcut of assuming symmetry as well, so that all the response thresholds are the same. It is true, however, that no such assumption need be made: a stationary equilibrium *must* be symmetric and yield the common response threshold described by (4.6). I omit the (simple) proof of this assertion.

4.4.3 Other Equilibria. Unfortunately, the uniqueness result for two-person Rubinstein bargaining no longer survives with three or more players. The argument, due to Herrero (1985) and Shaked (see Sutton (1986)) can be generalized to the full class of protocols we consider here. To write it down, consider any feasible allocation with nonnegative payoffs to each player. Define a *stage* to be any part of the game in which a proposal is about to be made (a proposal stage) or responded to (a response stage). In addition, define n special "rounds" as follows:

PHASE i: the proposer must give the entire cake to player i, and subsequently, all responders must accept this proposal.

---

[9]It is unnecessary to describe what happens if a later responder is planning to reject, as such a proposal will be rejected anyway and that is all that matters for our argument. Later, in the more general models we consider, this sort of consideration complicates the description of equilibrium strategy profiles (see, for instance, the proof of Proposition 5.1 in Chapter 5). But fortunately, such complications do not affect the main results in any way.

Now we connect the dots by describing when phase i must be started up. One way to do this is by developing the concept of "valid" and "invalid" stages (see [A] below), and by specifying how players must react to such stages (see [B] and [C] below).

[A] A stage is *valid* if it is a reaction to an invalid stage as specified by [B] or [C], or if it follows the actions specified at the very start of the game. Otherwise, it is *invalid*.

[B] Suppose the previous stage was valid. If this is a proposal stage and the last rejector was i, enter phase i. If this is a response stage, accept the going proposal.

[C] Otherwise, the previous stage is invalid. If the stage is a proposal stage, let i be the last rejector and start up phase "$i+1$", where $(i+1)$ stands for the index of the player after i (modulo n). If this is a response stage, the current responder j should reject the proposal if and only if she gets less than δ. (By [B], phase j is entered thereafter.)

This entire construction is similar to a penal code (Abreu (1988)) and can be employed to support various outcomes as equilibria. I claim that this penal code is itself an equilibrium whenever

(4.7) $$\delta > \frac{1}{n-1}.$$

To see this, we only need to check the unprofitability of one-shot deviations. First suppose the previous stage is valid. If a proposer does not abide by [B], then by [A], he starts an invalid stage. Because (4.7) holds, *some* responder must get less than δ under the deviant proposal, and so our deviant proposer must subsequently get zero. Similarly, if a responder does not accept the proposal from a valid proposal stage, then by [A], she starts an invalid phase and using [C], she will subsequently get zero. So it is a best response to follow [B] if the previous stage is valid..

Now for previous stages that are invalid. If the current stage is a proposer stage, a proposer will abide by the prescription in [B] for the same reason as in the previous paragraph. In a response stage, consider a responder who gets more than δ. If she rejects, then she starts an invalid phase and will surely get zero thereafter. If the responder gets less than δ, then by accepting she can get no more than what she is being currently offered.[10] By rejecting, she gets 1 after a lag discounted by δ. So it pays to abide by the prescription.

Finally, the fact that a deviant gets zero means there will also be no deviation from the initial play, and the argument is complete.

Note the crucial important point about this construction: it only works when $n \geq 3$, because the inequality (4.7) must hold.

The constructed penal code is extremely strong (use it to give any deviator a payoff of zero). So it can be employed to support all sorts of outcomes, including those that throw away part — or all! — of the available cake.

This embarrassing multiplicity of equilibria is similar to what we get in the folk theorem for repeated games (Aumann and Shapley (1976), Rubinstein (1979), Fudenberg and Maskin (1986)).[11] While not exactly a justification, it does explain why we restrict ourselves to stationary equilibria in the next few chapters.

That isn't to say that we abandon history-dependence entirely. We return to this theme in Chapters 8–10, where we study bargaining with reversible commitments.

4.5 Baron–Ferejohn Bargaining

Baron and Ferejohn (1989) study another model of noncooperative bargaining. Once again, several individuals divide a cake of unit size. However, unanimous assent is not required.

Specifically, suppose that there is an odd number n of players. Players are called upon according to some given proposer protocol to make proposals. Responses are sequential as they are here, but in contrast to the setup studied throughout this book, only a *majority* of respondents need approve the proposal for it to be implemented.[12] To complete the description of the protocol, a fresh proposer is

---

[10]For if she accepts, the proposal will either be implemented or someone else is asked to reject, and she will get zero.

[11]Of course, our bargaining framework is not a repeated game and that theorem doesn't formally apply here.

[12]Baron and Ferejohn (1989) call this procedure the *closed rule*. In contrast, under an open rule, a proposal may be amended before voting takes place. This extension need not concern us here.

selected at random if the previous proposal has been defeated: thus $\mu = 1/n$.

At first sight this looks like a very different model from unanimity bargaining, and indeed in a conceptual sense it is. But note that the model of bargaining with majority approval is easy to embed in an model of *coalitional* bargaining with *unanimity*. Simply construct the characteristic function

$$v(S) = 1 \text{ if and only if } |S| > \frac{n}{2},$$

and use the unanimity protocol! What is altered is essentially a matter of interpretation: a proposal requiring a unanimous response is never made to a subcoalition S, but it's *as if* it is: the proposal is in fact made to the grand coalition, with the implicit strategic presumption that the "targeted" majority subgroup S will in fact approve it.

Nor surprisingly, this sort of model acts as a game-theoretic endorsement of the view that "minimal" winning coalitions often form, an argument first given formal expression by Riker (1962).

In short, bargaining models that require majority approval can be easily embedded in coalitional bargaining models in which subcoalitions have power. In this sense there is little loss of generality in studying unanimity games, *provided* we are general enough to accommodate subcoalitional worths.

4.6 Summary

In this chapter, we've introduced a noncooperative model of coalition formation. Individuals make proposals to coalitions of their choice, and a proposal must be unanimously accepted in order for the coalition to form. A coalition S can write binding agreements among its members, so that they can enjoy payoffs in $\mathbf{U}(S, \pi)$ for every ambient coalition structure π that forms. At the same time, interaction with other coalitions is fully nonbinding and noncooperative.

It is assumed that a coalition once formed cannot be dismantled or added to. Coalition formation is an irreversible commitment. (We consider reversible commitments in Chapters 8–10.)

Rubinstein–Ståhl bargaining fits into this framework. So does the Baron–Ferejohn model.

In the chapters that follow, we go deeper by studying special cases of our general framework, beginning with the case of symmetry.

Appendix

Proof of Proposition 4.1. We proceed by induction on the number of players. Suppose an equilibrium exists with finite rounds of bargaining for every game with less than n players. For the one-player model, this assumption is trivially satisfied.

In what follows, suppose that a single coalition forms and exits, and there there are still some active players left. With some abuse of terminology (and only for the purpose of this proof), call this a *subgame*. On each subgame is induced a new bargaining game in the obvious way.[13] By our induction hypothesis, an equilibrium (with finite rounds of bargaining) exists for each such subgame: fix an equilibrium strategy profile for the players of that subgame. We describe equilibrium strategies for all the remaining nodes in the larger game, and graft these on to the fixed strategies for the subgames.

By choosing a fresh proposer according to some distribution, our protocol assigns a unique (stochastic) continuation to the subgame after a coalition S has formed, regardless of how S came to be. In particular, by fixing our subgame equilibrium, we have effectively set a probability distribution β^S over all coalition structures π conditional on S forming (obviously $\beta^S(\pi) > 0$ only if $S \in \pi$), and we have also assigned subgame payoffs u_i^S to every remaining player $i \notin S$.

Now consider player i in the game at hand. She can make an acceptable proposal to some coalition that contains her, or she can make an unacceptable proposal to some other player.[14] Let Δ_i be the set of all probability distributions over coalitions S with $i \in S$ (the choice of each coalition to be interpreted as the making of an acceptable offer to that coalition) as well as individuals $j \neq i$ (the choice of each j to be interpreted as the making of an unacceptable offer to that individual). Define Δ to be the product over the Δ_i's.

Because $\mathbf{U}(S, \pi)$ is compact for all π and $S \in \pi$, the equilibrium payoff (if there is one) to every player is obviously bounded above by some finite

---

[13]If coalition S exits and players in N' are left (that is, $N = S \cup N'$), define a new partition function by $\mathbf{U}^S(T, \pi) = \mathbf{U}(T, S.\pi)$ for every subcoalition T of N', where $S.\pi$ stands for the coalition structure π with S appended to it.

[14]There is no loss of generality in assuming that no unacceptable proposals are made to larger coalitions.

number m. We may therefore restrict the feasible payoff profiles to lie in M, the cube in $\mathbb{R}^n_+$ with vertex 0 and length m.

Fix a vector $\alpha \in \Delta$, and a vector $\mathbf{m} \in M$, the latter to be interpreted below as the vector of expected discounted equilibrium payoffs that each player receives in the game, if she were to reject an offer.[15] The following options are available to i.

First, she can choose S with $i \in S$, and make a proposal $\mathbf{u}_S(\pi)$ (conditioned on each π with $S \in \pi$). This will be interpreted in the sequel as an acceptable proposal. Consider the problem:

$$(4.8) \qquad \max_{\mathbf{u}_S(\pi)} \sum_{\pi} \beta^S(\pi) u_i(\pi)$$

subject to the constraints

$$(4.9) \qquad \sum_{\pi} \beta^S(\pi) u_j(\pi) \geq m_j, \qquad \text{for all } j \in S, j \neq i,$$

$$(4.10) \qquad u_S(\pi) \in \mathbf{U}(S, \pi) \qquad \text{for each } \pi \text{ with } S \in \pi.$$

Denote by $g_i(S, \mathbf{m})$ the maximum value so attained. The assumed comprehensiveness of $\mathbf{U}$ guarantees that g_i a continuous function of $\mathbf{m}$. Also note that g_i is assuredly nonnegative: player i can always walk away on her own, leaving the bargaining to end in a finite number of rounds (by the induction hypothesis).

Second, i might make an unacceptable proposal to j.

Both these cases can be considered together in the following way. Recall that we are given some $(\mathbf{m}, \alpha) \in M \times \Delta$. For every i, construct the following system $\{V^j_i(\mathbf{m}, \alpha), w^j_i(\mathbf{m}, \alpha)\}_{j \in N}$:

$$(4.11) \qquad V^j_i(\mathbf{m}, \alpha) = B^j_i + \sum_{k \neq j} \alpha_j(k) w^k_i(\mathbf{m}, \alpha),$$

and

$$(4.12) \qquad w^j_i(\mathbf{m}, \alpha) = \delta \sum_{k \in N} \rho(j, k) V^k_i(\mathbf{m}, \alpha),$$

where $\rho(j, k)$ is the protocol probability that k will be asked to make a proposal following a rejection by player j,[16] and where

$$(4.13) \qquad B^i_i \equiv \sum_{S} \alpha_i(S) g_i(S, \mathbf{m}),$$

---

[15] This is well-defined, as we allow subsequent play to depend on the identity of the last rejector, but no more.

[16] We've actually made specific assumptions on the protocol but those won't be needed here.

and for $j \neq i$,

$$(4.14) \qquad B_i^j \equiv m_i \left[\sum_{S:i\in S} \alpha_j(S) \right] + \sum_{S:i\notin S} \alpha_j(S) u_i^S.$$

We may interpret V_i^j in (4.11) as the expected payoff to i when player j proposes. The right hand side of that equation has two terms. The first, B_i^j, can be seen from (4.13) and (4.14) to be the expected amount that i gets when j makes an acceptable proposal. The second term captures the payoff to i when j makes an unacceptable proposal to one of various players. This interpretation of w_i^k is indeed justified on examination of (4.12).

In particular, $V_i^i(\mathbf{m}, \alpha)$ is to be interpreted as i's expected payoff when i herself is the proposer.

It is every easy to see that the system of linear equations (4.11) and (4.12) has a unique solution for $\{V_i^j(\mathbf{m}, \alpha), w_i^j(\mathbf{m}, \alpha)\}_{j\in N}$, which is continuous in $(\mathbf{m}, \alpha)$; we omit the details.

Now define a function on $M \times \Delta \times \Delta_i$ by

$$(4.15) \qquad v_i(\mathbf{m}, \alpha, \alpha_i') \equiv \sum_S \alpha_i'(S) g_i(S, \mathbf{m}) + \sum_{j \neq i} \alpha_i'(\{j\}) w_i^j(\mathbf{m}, \alpha),$$

and maximize this function with respect to $\alpha_i' \in \Delta_i$.

Let $\phi_i^1(\mathbf{m}, \alpha)$ denote the set of maximizers of this problem. It is easy to see that $\phi_i^1(\mathbf{m}, \alpha)$ is a nonempty, convex-valued, upper hemicontinuous correspondence.

Next, define $\phi_i^2(\mathbf{m}, \alpha)$ to be the maximum value of this problem. It is easy to see that $\phi_i^2(\mathbf{m}, \alpha)$ is continuous and nonnegative. Consequently,

$$\psi_i(\mathbf{m}, \alpha) = \delta \left[\rho(i, i)\phi_i^2(\mathbf{m}, \alpha) + \sum_{j \neq i} \rho(i, j) V_i^j(\mathbf{m}, \alpha) \right]$$

is also continuous in $(\mathbf{m}, \alpha)$. Also, it is obvious that $\prod_i \psi_i$ maps from $M \times \Delta$ into M. Therefore the correspondence

$$\prod_i \psi_i \times \prod_i \phi_i^1 : M \times \Delta \mapsto M \times \Delta$$

satisfies all the conditions of Kakutani's fixed point theorem and has a fixed point $(\overline{\mathbf{m}}, \overline{\alpha})$.

We shall now use this fixed point to construct an equilibrium. Let σ denote the strategy profile such that:

(i) When the player set is N, player i as a proposer makes proposals according to $\overline{\alpha}_i$. To every coalition S containing i with $\overline{\alpha}_i(S) > 0$, she proposes $\mathbf{u}_S(\pi)$ which solves the problem defined by (4.8), (4.9) and (4.10). To every $j \neq i$ such that $\overline{\alpha}_i(\{j\}) > 0$, she makes an unacceptable proposal: offering in expectation any amount strictly less than $\overline{m}_j$ (possibly negative). By construction of the fixed point, this yields player i a discounted payoff of precisely $\overline{m}_i$, conditional on rejecting an offer.

(ii) Suppose the player set is N, and that player $i \in S$ is responding to a proposal $\{\mathbf{u}_S(\pi)\}$. Proceed inductively on the number of respondents to follow player i. If there are none, i accepts if and only if

(4.16) $$\sum_\pi \beta^S(\pi) y_i(S, \pi) \geq \overline{m}_i.$$

Inductively, suppose we have computed the accept-reject decisions of k respondents to follow i. If there are no rejectors following i, use the rule in (4.16) again. If there is a first rejector j following i, her decision will now depend on the present value of her payoff resulting from j rejecting the offer. In fact this value is precisely $w_i^j(\mathbf{m}, \overline{\alpha})$; see (4.12). We therefore require that player i accept the proposal if and only if

$$w_i^j(\mathbf{m}, \overline{\alpha}) \geq \overline{m}_i.$$

(iii) At all other player sets we append the equilibrium strategy profiles chosen in the inductive step at the very beginning of this proof.

It is easy to see that a strategy profile satisfying (i)–(iii) is a stationary equilibrium. First consider part (i), and any player i. By construction, α_i is chosen to maximize (4.15), so as proposer no profitable one-shot deviation is possible. Next, observe that by construction, $\overline{m}_i$ is player i's payoff under the given strategy profile from rejecting an offer. The (one-shot) optimality of the responses prescribed in (ii) follows as an immediate consequence. Item (iii) requires no additional discussion.

So no one-shot deviation is profitable, and we've therefore found a stationary Markovian equilibrium. Moreover, this equilibrium has nonnegative payoffs, so bargaining must end in finite time.   □

Remark. Our method of proof identifies a stationary Markovian equilibrium in which mixing may be necessary, but in which the *only* source of mixing is in the (possibly) probabilistic choice of a coalition by each proposer. Ray and Vohra (1999) show that this degree of mixing is generally necessary for the existence of equilibrium.

It is also of interest to ask whether our theorem extends without any modification to games with "fully nontransferable" segments. Comprehensiveness fails for such games, although it is true that comprehensive games can approximate lack of transferability arbitrarily closely.

But these are largely technical matters. In the cases of interest to follow, existence will not be a concern.

CHAPTER 5

Irreversible Agreements: Symmetric Games

An analysis of the general model introduced in the previous chapter is complicated, though we hope to make some serious progress. We begin by studying *symmetric* partition functions with transferable utility. Such functions have the property that the worth of a coalition depends only on the *number* of individuals in that coalition and the ambient "numerical coalition structure" — the ambient coalition structure described only by the vector of membership sizes.

Though we assume both symmetry and transferability, symmetry is really the main assumption at work. Under some mild qualifications, our analysis can be easily extended to symmetric sets of nontransferable payoffs (see Section 5.4 for a brief discussion).

Observe that the symmetry postulate places all individuals, *ex ante*, in a similar position. But this is emphatically not to say that all individuals must always receive the same payoffs: only that coalitions of the same size *and* in the same ambient numerical structure must enjoy the same coalitional worth. So *ex post*, depending on the particular configuration that different individuals find themselves in, there may be substantial variation in payoffs.

While we go beyond the symmetry assumption to a substantial degree in Chapter 7, it is worth noting that several applications fit into this category. We discuss two of them in Chapter 6.

The main results of this chapter involve the unearthing of a *particular* class of coalition structures, with the property that such a class is predicted by a broad class of equilibria, within a general class of bargaining protocols. Under some conditions this prediction is unique.

In summary, given the assumptions that commitments are irreversible and that the underlying game is symmetric, we obtain a remarkably sharp prediction of equilibrium outcomes. Before we get into the analysis, it may be useful to describe this prediction informally.

Roughly speaking, we identify a particular coalition structure created by *recursively maximizing average worth*. That is, at every stage of the game, the size of the coalition that forms next will be given by the coalitional size that maximizes the average worth across all coalitions. But this description is incomplete, because the "average worth" of a coalition is contextual when there are externalities: it depends on the ambient structure within which the coalition is embedded. This is where the word "recursive" comes in: such a structure is calculated by supposing that average worth maximization will be carried out in all later stages as well.

This apparently circular bout of reasoning is actually quite linear, because we can proceed by recursion on the number of stages left in the game. By starting with the simplest environments in which only a single active player is left, we can work backwards to generate a full coalition structure that satisfies recursive average worth maximization. As we shall see in the examples, the structure can be quite complicated — after all, apart from symmetry we impose very few assumptions — but it is generated by the elementary conditions of average worth maximization and is thus eminently applicable.

A corollary of this characterization is that equilibrium coalition structures are not, in general, efficient. The latter necessitates the maximization of *total* worths, while our prediction maximizes average worth. Later, we discuss how general this particular property is.

5.1 Symmetric Partition Functions

A partition function is *symmetric* if the worth of a particular coalition in a given partition depends *only* on the number of individuals in each coalition in that partition.

With a little abuse of notation, the transferable worth $v(S, \pi)$ of a coalition S in π can be written as $v(s, \mathbf{n})$, where s is the size of S

and **n** is the collection of coalitional sizes in π. We will call **n** a *numerical coalition structure*, and will occasionally use the notation $\mathbf{n}(\pi)$ to connect **n** to the underlying structure π.

It will also be useful to describe the partition function as a vector of worths, one for each coalition in the partition: we will use the notation $\mathbf{v}(\pi)$ or $\mathbf{v}(\mathbf{n})$ to do this.

5.2 An Algorithm

In this section, we recursively construct the coalition structure that will be of central interest. There is no game theory in the algorithm that follows and you can look at it simply as a mathematical construction. What will give it meaning are the propositions that will link the equilibria of our game to the structures generated by this algorithm.

In what follows, objects such as **n** will refer not just to a numerical coalition structure for the full game at hand, but equally to numerical *substructures*, collections of positive integers that add up to any number strictly less than the total number of players. For any such substructure $\mathbf{n} = (s_1, \ldots, s_m)$, define $K(\mathbf{n}) \equiv \sum_j s_j$; then $K(\mathbf{n}) < n$. Use the notation ϕ to refer to the "zero-dimensional" or null substructure containing no entries, and set $K(\phi) = 0$. Let $\mathcal{F}$ be the family of all substructures (including ϕ). While, as I've said already, there is no game theory in the analysis to follow, it is useful to interpret a substructure as a collection of coalitions that has "already formed" in a subgame.

We are going to construct a rule $t(\mathbf{n})$ that assigns to each member of $\mathcal{F}$ a positive integer. Continuing in our interpretative vein: given that the substructure **n** has already formed, $t(\mathbf{n})$ is the size of the coalition that forms *next*. By applying this rule repeatedly starting from ϕ, we will generate a particular numerical coalition structure, to be called $\mathbf{n}^*$.

STEP 1. For all **n** such that $K(\mathbf{n}) = n - 1$, define $t(\mathbf{n}) \equiv 1$.

STEP 2. Recursively, suppose that we have defined $t(\mathbf{n})$ for all substructures **n** with $K(\mathbf{n}) \geq m + 1$ for some $m \geq 0$. For any such **n**, define

$$c(\mathbf{n}) \equiv \mathbf{n}.t(\mathbf{n}).t(\mathbf{n}.t(\mathbf{n})) \ldots ,$$

where the notation $\mathbf{n}.t_1.\ldots.t_k$ simply refers to the numerical coalition structure obtained by concatenating $\mathbf{n}$ with the integers $t_1,\ldots,t_k$.

STEP 3. For any $\mathbf{n}$ such that $K(\mathbf{n}) = m$, define $t(\mathbf{n})$ to be the *largest* integer in $\{1,\ldots,n-m\}$ that maximizes the expression

(5.1)
$$\frac{v(t,c(\mathbf{n}.t))}{t}.$$

STEP 4. Complete this recursive definition so that t is now defined on all of $\mathcal{F}$. Define a numerical coalition structure (for the entire set of players) by

$$\mathbf{n}^* \equiv c(\phi).$$

This completes the description of the algorithm. There will soon be plenty of examples to give us a feel for $\mathbf{n}^*$. But first we must show that $\mathbf{n}^*$ is deserving of special attention.

5.3 Connecting the Algorithm to Equilibria

This section proceeds in steps. We first show that the algorithmic structure $\mathbf{n}^*$ is predicted by a "reasonable" class of equilibria, which we call *standard equilibrium*. We then to proceed to rule out other equilibria. These exercises will require some conditions on the partition function (so far we have imposed none, except for symmetry).

5.3.1 Standard Equilibrium. An equilibrium is *standard* if every proposer at every stage makes an acceptable proposal with positive probability.

The terminology "standard" is deliberately chosen to suggest that such equilibria are reasonable. There is no incompleteness of information, so why make an unacceptable proposal that is guaranteed to cause delays? Surely, there should be no real advantage to delaying one's inclusion in a formed coalition?

These questions (rhetorical though they may be) are sensible. In general, equilibrium proposals *will* be acceptable; in the examples of economic interest that we consider later, this is certainly the case. But it is also important to note that — barring symmetry — we have made absolutely no assumptions on the shape of the partition function. In certain situations, some equilibria need not be standard, and in some cases, standard equilibria need not exist at all.

At the risk of appearing entirely perverse, I am now going to present an example with "nonstandard" equilibria. The reader must keep in mind that this example is not intended to cast doubt on the reasonableness of standard equilibrium. However, if we are going to impose further conditions on the game, it is useful to illustrate the need for such restrictions, even though they may come across as mild restrictions anyway.

Here is the example:

EXAMPLE 5.1. *There are five players. The partition function is given by*

$$v(4,1) = (6,2), \qquad v(3,2) = (3,8), \qquad v(2,1,1,1) = (0.1,3,3,3),$$

$$v(3,1,1) = (10,0,0), \qquad v(n) = 0 \quad \text{for all other } n.$$

The first rejector for any proposal makes the next proposal.

I've deliberately put grand-coalitional worth to 0 to cut down on the number of cases (a similar example can easily be constructed if the game is required to be grand-coalition superadditive).

Proceeding informally for a moment, consider possible equilibria in which a coalition of four players forms, and one player is left out. Notice that at least one player in the four-player coalition must receive no more than 1.5, whereas by being the lone outsider she can do. Therefore it pays for at least one player to stand apart while the remaining four players form a coalition. One way to do this is to have that player unilaterally and irreversibly exit, leaving the other four players to their devices. But in that case the four-player coalition will *not* form. Instead, three of the remaining players will form a coalition, leaving the original player with zero.

The only way for a player to avail of the high outsider payoff is to "wait" until the four-person coalition has formed, and she achieves this by making unacceptable offers. More formally, for all discount factors sufficiently close to unity, there is an equilibrium with numerical coalition structure $(4,1)$ in which one player makes an unacceptable proposal (in the presence of all five players) and the other four make acceptable proposals to each other. This equilibrium is obviously not standard. Under it, the "intransigent" player receives 2δ whenever it is his turn to propose, and all players are active. The others receive only $\frac{6}{1+3\delta}$ in their role as proposer. We leave the details of equilibrium construction to the reader.

There is *another* equilibrium — again, not standard — with coalition structure $(3, 2)$. It is constructed as follows. Players 1, 2 and 3 make acceptable offers to each other and the other two make unacceptable offers to player 1. Let $\bar{x}_i$, the equilibrium payoff to i if i starts the game, be defined as

$$\bar{x}_i = \frac{3}{1 + 2\delta} \quad \text{for } i = 1, 2, 3,$$

$$\bar{x}_j = \frac{8\delta}{1 + \delta} \quad \text{for } j = 4, 5.$$

For δ close to 1, players 1, 2 and 3 get approximately 1 while players 4 and 5 get approximately 4. Clearly, none of the first three players can do better by including player 4 or 5, since $v(4, 1) = (6, 2)$. Nor, given the strategies of the others, can players 1–3 gain by making an unacceptable proposal. Similarly, it is easy to see that players 4 and 5 do not have a profitable deviation. Therefore, the above strategies (together with obvious specifications for non-equilibrium subgames) constitute an equilibrium.

There may still be more equilibria, but what can certainly be shown is that *none of these can be standard* under the rejector-proposes protocol. Having Proposition 5.2 in hand helps us to establish this very quickly, so we postpone a more detailed discussion of this assertion until the statement of that proposition.

Nonstandard equilibria are characterized by the making of absurd offers which the proposer knows will be rejected.[1] Why would such offers ever be made? The answer is that a proposer may wish to pass the buck to another player, and benefit from possibly higher payoffs in some subgame. Assuming that players within a coalition divide their algorithmic worth roughly equally (we will need to prove this of course), this suggests that the problem goes away if average algorithmic worth $a(\mathbf{n})$ declines "as the algorithm proceeds". This is the route we explore below.

---

[1]One might argue that the making of *unacceptable* offers is merely an artifact of the assumption that a player must include herself in any proposal. But the problem simply reappears in a different guise: players may make acceptable offers, but they will do so to coalitions that don't include themselves, preferring instead to form a coalition later in the proceedings.

5.3.2 Nonincreasing Average Worths. For each numerical substructure $\mathbf{n}$, define

(5.2)
$$a(\mathbf{n}) \equiv \frac{v(t(\mathbf{n}), c(\mathbf{n}.t(\mathbf{n})))}{t(\mathbf{n})}.$$

The numbers $a(\mathbf{n})$ can, of course, be directly computed from the primitives of the model. These are the *average worths* of the algorithm at each "stage $\mathbf{n}$".

Say that algorithmic average worth is *nonincreasing* if

$$a(\mathbf{n}) \geq a(\mathbf{n}.t(\mathbf{n}))$$

for every substructure $\mathbf{n}$ such that $\mathbf{n}.t(\mathbf{n})$ is also a substructure.

Nonincreasing average worth — or NAW henceforth — is our first real restriction on the partition function (apart from symmetry). It may be worth pausing a moment to examine it further.

Note, first, that NAW has bite *only* for partition functions. For all characteristic functions, in which the worth of a coalition depends only on the coalition itself, the condition must be trivially satisfied. The reason is simple. Our algorithm involves the stepwise maximization of average worth, setting each maximizing coalition aside as the algorithm proceeds. If there are no externalities across coalitions, such a process must result in a sequence of (maximal) average worths that can never increase; for if they did, such coalitions would have been chosen *earlier* in the algorithm, not *later*.

This simple observation also assures us that NAW is not connected to other well-known features such as superadditivity. *All* (symmetric) characteristic functions satisfy NAW.

Whether or not NAW applies more generally (i.e., when externalities are present) is a less transparent question, and the answer will largely depend on the application at hand. If personal experience is of any help, I haven't come across an interesting economic or political application where NAW isn't satisfied, though I certainly cannot rule out the possibility. For instance, the two examples to be examined in detail in Chapter 2 both satisfy NAW. But it is certainly possible to write down a mathematical example in which NAW fails: Example 5.1 is one such instance.

5.3.3 More on Protocols. As described in the previous chapter, we consider a fairly broad class of protocols in which the first rejector of a proposal has the option to exit the game, but conditional on staying she gets to be the next proposer with some probability. If this probability is always one, we have the *rejector-proposes protocol*, familiar initially from the alternating-offer bargaining model of Rubinstein (1982), and used in a large literature on bargaining. If this probability is equal across all active players, the rejector has no particular claim on making a fresh proposal (as in Baron and Ferejohn (1989)), and if this probability is close to zero, the protocol is actually biased *against* the rejector.[2] We've assumed, however, that in all such cases the first rejector has the option to unilaterally (and irreversibly) exit the game.

We will presently discuss the sense in which our results are robust across this wide variety of protocols, but it will be convenient to begin in a somewhat narrower class. Say that a protocol is *rejector-friendly* if the rejector gets to counterpropose with better than even probability, and if this probability does not decline as the number s of active players decreases: $\rho(s) > 1/2$, and is nonincreasing in s. Our results are initially restricted to this subclass, but we then discuss (see Section 5.3.4.4 below) how they might extend to a broader context.

5.3.4 NAW, Standard Equilibria and Equilibrium Coalition Structure.
The following pair of propositions relates NAW to standard equilibria. Proposition 5.1 unearths a particular coalition structure predicted by all rejector-friendly protocols, provided that bargaining frictions are close to zero. Proposition 5.2 shows that this structure is the *only* one *common* to standard equilibria as we move over all rejector-friendly protocols. We begin with precise statements of the propositions and then proceed to a more informal discussion.

PROPOSITION 5.1. *Under NAW, for every rejector-friendly protocol, there exists a standard equilibrium yielding the algorithmic coalition structure* $\mathbf{n}^*$, *provided that the discount factor is sufficiently close to one.*

PROPOSITION 5.2. *Provided that the discount factor is sufficiently close to one,* $\mathbf{n}^*$ *is the only coalition structure under standard equilibrium in the rejector-proposes protocol, and is therefore the only structure common to standard equilibrium in all protocols.*

---

[2]No restriction is imposed on the choice of a fresh proposer for the remaining active players, following an *accepted* proposal. But see Section 7.7 of Chapter 7 for more discussion in the general, asymmetric case.

There are several aspects of these propositions that require further discussion.

5.3.4.1 Why n*? First, the reason why n^* is singled out is that the equilibrium behavior we identify is connected closely to *equal* division of the available worth as bargaining frictions go to zero. One way to see this is to simply recall the two-person Rubinstein–Ståhl bargaining model: there, as the common discount factor converges to one, the bargaining outcome converges to equal division of the available surplus. Indeed, as Section 4.4 in Chapter 4 shows, such convergence to equal division holds over a variety of protocols, which includes all the protocols in the statements of the propositions. Equal division is, therefore, a very natural focal point in the symmetric case (when two identical players bargain over a prize, should we expect anything else?[3]), and our result may be viewed as a transplant of this property it to the far broader framework of this chapter. The algorithm supplies the correct generalization of "equal division" in our extended framework.

5.3.4.2 The Role of NAW. The intuitive appeal of n^* notwithstanding, we must be careful here. Obviously, matters are not *that* simple, as Example 5.1 already reveals. The far more nuanced strategic structure of our model requires additional conditions for the equal division intuition to carry through. NAW provides such conditions. To see why these are needed in general, return to Example 5.1. Proposition 5.2 (together with some additional calculations) can be used to demonstrate that there is no standard equilibrium in that example, as $\delta \rightarrow 1$. Here is the argument.

Suppose, on the contrary, that such an equilibrium exists along a sequence of discount factors tending to unity. Apply the algorithm to the example to check that $n^* = (4, 1)$. Therefore — recalling that rejectors propose in that example — Proposition 5.2 tells us that a coalition of four individuals must form first along the equilibrium path. It is easy to calculate that a proposer receives $\frac{6}{1+3\delta}$, and a responder receives $\frac{6\delta}{1+3\delta}$.

---

[3]It is true, for instance, that if discount factors are heterogeneous even as they all converge to unity, unequal division might occur. I do not take this case seriously as a possible reason for unequal division, and therefore presume that the discount factor is common to all players. See footnote 8 in Chapter 4 for more discussion.

Fix any $\delta \geq \delta^*$ such that $\delta[0.5 + 6\delta] > 6$. Now observe that there is *some* pair of individuals i and j such that if it is j's turn to propose, an offer is made to i with probability *no more* than 3/4. If individual i deviates by making an unacceptable offer to j, then the present value of i's payoff is bounded below by $\delta[\frac{3}{4}\frac{6\delta}{1+3\delta} + \frac{1}{4}2]$, while by sticking to equilibrium policy, he obtains $\frac{6}{1+3\delta}$. Comparing these two expressions under the given restriction on δ, it can easily be checked that a deviation is profitable. This completes the argument.

A sharper query: is NAW logically necessary for our Propositions, or might a weaker condition (which nevertheless excludes Example 5.1) suffice? It can be shown that NAW is actually necessary for the existence of a *pure strategy* standard equilibrium, as long as the rejector-proposes protocol is included in the class over which existence is demanded. See Ray and Vohra (1999, Theorem 3.3).

5.3.4.3 Other Structures in Standard Equilibrium. There is a second reason for caution. Proposition 5.2 makes it clear that $\mathbf{n}^*$ is the unique coalition structure that is common to all *every* rejector-friendly protocol. As the statement of that proposition makes explicit, this is so because the rejector-*proposes* protocol admits no other outcome (in standard equilibrium). But the other protocols might. To see this, consider the following example:

EXAMPLE 5.2. *There are four players. The partition function is given by*

$$v(3,1) = (22,0), \quad v(2,1,1) = (16,1,1), \quad v(\mathbf{n}) = 0 \text{ for all other } \mathbf{n}.$$

The first rejector for any proposal can exit, as described in the text, or gets to make the next proposal with probability 7/12.

Applying the algorithm is seen to yield the coalition structure $\mathbf{n}^* = \{2,1,1\}$. It is very easy to check that NAW holds. Indeed, as asserted in the proposition, there is a standard equilibrium (for low bargaining frictions) that generates $\mathbf{n}^*$. But there is *another* standard equilibrium as well, one that generates the coalition structure $\{3,1\}$. I construct this equilibrium by studying a strategy profile in which — at the initial stage with all players active — acceptable proposals are made to a three-person coalition, and the rejector of a proposal is excluded from the next proposal if she does not get to make that proposal herself.

It is instructive to carry out this construction. Denote by x the equilibrium payoff to a proposer and by m the equilibrium payoff to a responder, conditional on rejecting an offer. Suppose that an acceptable offer is made to a three-person coalition (we shall verify this below). Then

$$x = 22 - 2m \quad \text{and} \quad m = \delta \frac{7}{12} x,$$

where the second equality incorporates the feature that a rejector is excluded from a counterproposal unless it is made by the rejector herself.

For δ is close enough to 1, this shows that m is approximately 5.92. Faced with these response thresholds, we must now verify that a proposer will indeed prefer to propose to a three-person coalition, obtaining $22 - 2m$ rather than proposing to a two-person coalition and obtaining $16 - m$. Performing the necessary sums, it is trivial to see that the three-person coalition will indeed be preferred. This completes an informal demonstration that every standard equilibrium need not yield $\mathbf{n}^*$ even if bargaining frictions are close to zero.

The reason we nevertheless manage to find *some* standard equilibrium that does yield $\mathbf{n}^*$ is that an equilibrium strategy profile which *includes*, rather than excludes, the rejector of the last proposal in every counterproposal can always be constructed. Such an equilibrium must generate equal division in every formed coalition, and we are back to our predicted structure $\mathbf{n}^*$. (Notice that in the construction of the previous paragraph, equal division is *not* obtained.) You can easily construct such an equilibrium in this example, or if you prefer, study the proof of Proposition 5.1 which achieves this task in greater generality.

5.3.4.4 Robustness to an Even Wider Class of Protocols. An uncomfortably familiar feature of bargaining models is that their predictions are often sensitive to the finer points of procedure — to the *protocol*, in the language of this chapter. This is why I have taken care to present results that cover a broad class of protocols. One might worry, though, that the class isn't broad enough. For instance, the two propositions in this sections have been stated for the class of rejector-friendly protocols, procedures in which the rejector has quite a bit of power. Rejector-friendly protocols exclude, for instance, the case of a uniform proposer protocol, one in which a new proposer is chosen with uniform probability over the set of active agents whenever a going proposal has been rejected. Do our results extend to such cases: is $\mathbf{n}^*$ still predicted by (some) standard equilibrium for this and similar protocols?

To answer this question, recall our description of the n-person Rubinstein–Ståhl bargaining model with Markovian strategies. In that model, we permit a wide range of protocols which include the class studied in propositions 5.1 and 5.2, the "uniform protocol" mentioned in the previous paragraph as well as a variety of other procedures. Yet equal division is always implemented as the common discount factor converges to 1. This suggests that it should be possible to extend our results to protocols in the wider class.

A careful study of the proof of Proposition 5.1 reveals the difficulty. (The more hurried reader is invited to read the remarks following the proofs of Lemmas 5.2, 5.5, and Proposition 5.1.) Recall our discussion of the NAW condition. It pushes the proposer to make acceptable proposals in equilibrium, by lowering average worths in future stages. An important part of our formal proof deals with the translation of the NAW restriction to the discounted case at hand, at least when discount factors are close to unity. This formal argument becomes extremely complex when NAW holds with *equality* in one or more stages. If I were to drop equality from NAW, requiring instead that

(5.3) $$a(\mathbf{n}) > a(\mathbf{n}.t(\mathbf{n}))$$

for every substructure $\mathbf{n}$, then it is possible to significantly widen the class of protocols. In short, it is possible to prove

PROPOSITION 5.3. *If NAW holds in the stricter sense described by (5.3), Propositions 5.1 and 5.2 continue to hold for every protocol in which the rejector, apart from unilateral exit rights, has some positive probability of making a counterproposal.*

A second robustness check on protocols has to do with the unilateral exit rights we've conferred on any rejector. Recall that this simply means that a rejector can form a "one-person coalition" and exit if she so pleases, or she can stay on and submit herself to the vagaries of a new choice of proposer (which might be herself or someone else). Are the propositions robust to the dropping of such rights?

To understand this question, recall how we construct the equilibrium in which equal division is "implemented" at every stage. We do so by including the last rejector in every counterproposal (recall the discussion following Example 5.2). But if the optimal coalition size is 1 (i.e., if $t(\mathbf{n}) = 1$ at any stage $\mathbf{n}$) this is impossible to do unless the rejector herself gets to make the counterproposal. This snag might cause a failure of the algorithm,[4] but giving a rejector unilateral exit rights gets around the problem. Notice that if singleton coalitions are never chosen by the algorithm whenever there are two or more active players, this problem disappears and our results are robust to an even broader class of protocols, in which it does not matter whether or not the rejector is given exit rights.

5.3.5 On the Uniqueness of $\mathbf{n}^*$.

Proposition 5.2 argues that $\mathbf{n}^*$ is the only equilibrium coalition structure that is common to standard equilibria across all rejector-friendly protocols, provided that bargaining frictions are small. That proposition also makes clear that such a singling-out is achieved by the rejector-proposes protocol, which admits no other standard equilibrium. As we have seen (see, for instance, Example 5.2), other protocols do permit standard equilibria to generate alternative coalition structures, but $\mathbf{n}^*$ is always among the predicted structures. In this sense $\mathbf{n}^*$ appears to be a focal prediction.

But this finding is limited by our restriction to standard equilibrium. It is entirely possible that there are other (Markovian) equilibria which generate very different coalition structures, and that they persist even under the rejector-proposes protocol. To see this, consider the following variation on Example 5.1:

EXAMPLE 5.3. *There are five players. The partition function is given by*

$$\mathbf{v}(4,1) = (6,1), \qquad \mathbf{v}(3,2) = (3,8), \qquad \mathbf{v}(2,1,1,1) = (0.1,3,3,3),$$

$$\mathbf{v}(3,1,1) = (10,0,0), \qquad \mathbf{v}(\mathbf{n}) = 0 \text{ for all other } \mathbf{n}.$$

The first rejector for any proposal makes the next proposal.

The only change relative to Example 5.1 is the specification of $\mathbf{v}(4,1)$. But this change guarantees that NAW is satisfied. Proposition 5.1 tells us that a standard equilibrium must make an appearance, and indeed one does when bargaining frictions are small, yielding the

---

[4]For instance, suppose that $n = 10$, $\mathbf{v}(1,9) = (2,1)$, $\mathbf{v}(10) = 10$, and $\mathbf{v}(\mathbf{n}) = 0$ for other structures $\mathbf{n}$. Then $t(\emptyset) = 1$, so the proposition would state that a standard equilibrium exists with a singleton coalition forming at the very first stage. However, think of the uniform protocol in which a new proposer is chosen uniformly following any rejection, and no rejector has exit rights. Then it is possible to check that no such standard equilibrium exists: the proposer will wish to propose the grand coalition.

coalition structure $(4, 1)$. However, the asymmetric equilibrium of Example 5.1, with the coalition structure $(3, 2)$, persists here as well. What is more, this equilibrium is in no hurry to go away as we move over our range of protocols. This shows that NAW isn't enough to single out $\mathbf{n}^*$ (over different protocols), unless we restrict ourselves to standard equilibria.

Such a restriction may be perfectly fine. Perhaps we're not interested in asymmetric equilibria. On the other hand, it might be reassuring to be able to tighten our predictions further. The purpose of this section is to show that under conditions somewhat stronger than NAW, our algorithmic structure $\mathbf{n}^*$ is the *only* predicted equilibrium coalition structure common to all rejector-friendly protocols (with low bargaining frictions), *whether or not we restrict ourselves to standard equilibria*. This is an even stronger prediction that pushes us closer still to accepting $\mathbf{n}^*$ as a solution for symmetric games. Once again, the rejector-proposes protocol accomplishes the narrowing of equilibrium outcomes: we will establish uniqueness under this protocol.

But first let us see what the stronger condition is. To this end, define for each $\mathbf{n} \in \mathcal{F}$, and each positive integer $\ell \in \{1, \ldots, n - K(\mathbf{n})\}$,

$$(5.4) \qquad \tau_\ell(\mathbf{n}) \equiv \arg \max_{t \in \{1, \ldots, \ell\}} \frac{v(t, c(\mathbf{n}.t))}{t}.$$

In words, $t \in \tau_\ell(\mathbf{n})$ solves the same maximization problem as described in the algorithm, except that maximum size is restricted by ℓ.[5]

Now consider a strengthening of NAW. Say that algorithmic average worth is *strongly nonincreasing* ("strongly nonincreasing average worth", or SNAW) if for each $\mathbf{n} \in \mathcal{F}$ and each positive integer $\ell \in \{1, \ldots, n - K(\mathbf{n})\}$,

$$(5.5) \qquad a(\mathbf{n}) \geq a(\mathbf{n}.t) \qquad \text{for all } t \in \tau_\ell(\mathbf{n}).$$

This condition is stronger than NAW, which requires only that average worth weakly decline as we progress along the algorithm, step by step. More precisely, think of maximizing average worth at a stage in which the partial structure $\mathbf{n}$ has already been removed. Suppose it is maximized at the integer t. Then NAW only asks that average worth not be increasing at the next step of the algorithm,

---

[5]Because of possible nonconvexities, this maximum restriction can be binding even if $\ell \notin \tau_\ell(\mathbf{n})$. Of course, $t(\mathbf{n}) \in \tau_\ell(\mathbf{n})$ whenever $\ell \geq t(\mathbf{n})$.

when the removed partial structure is given by **n**.*t*. SNAW asks for more. It requires that this condition hold not just for the maximizing integer *t*, but for all "constrained maximizers" *t'* obtained by solving the average worth maximization problem over integers up to some upper bound. Because the unconstrained maximizer *t* is also a constrained maximizer for some large enough bound, SNAW is a genuine strengthening of NAW.

The utility of introducing this condition is brought out in

PROPOSITION 5.4. *Under SNAW, provided that discount factors are sufficiently close to one, the only equilibrium coalition structure common to all rejector-friendly protocols is* **n***.

Proposition 5.4 provides a sharper prediction (under stronger assumptions). It underlines the focal nature of the coalition structure **n*** as an equilibrium prediction for symmetric partition functions. Once again, it should be pointed out — and the proof of the proposition will make this very clear — that the uniqueness result is really driven by the presence of the "rejector-proposes" protocol, under which **n*** is the unique prediction, given SNAW. Of course, this is not to shun the usefulness of considering other protocols, in which **n*** is always *an* equilibrium prediction.

Like NAW, SNAW is trivially satisfied for all characteristic functions. Whether SNAW holds more abundantly in economic or political applications with externalities remains to be seen.[6] We will explicitly verify SNAW in the applications that we consider later in this book; see Chapter 6.

5.4 A Remark on Nontransferable Payoffs

Within the class of symmetric games, it isn't hard to incorporate arbitrary degrees of nontransferability. Suppose that we retain all the symmetry assumptions, but replace the TU worth $v(S, \pi)$ by some *symmetric* set of payoffs $\mathbf{U}(S, \pi)$. Nothing of substance will change as long as we are willing to assume that each such set is convex.

While the details are already complex (as the proofs in the next section will reveal) and dropping transferability certainly does not help, it is easy to obtain an intuition of why the same arguments

---

[6]Of course, it does fail in Example 5.3, which was artificially concocted to show why something like SNAW is needed.

go through. Average worth will now need to be replaced by the symmetric utility obtained "along the diagonal" for each $\mathbf{U}(S, \pi)$, and the same algorithm may be written down with average worth replaced by this symmetric utility. It is easy enough to see why convexity may be needed for such an argument. There is no guarantee otherwise that "equal division" will be followed within all coalitions, and our techniques cease to apply.

Section 14.8 of Chapter 14 contains further discussion on the NTU case.

5.5 Proofs

In several chapters starting with this one, I will collect formal derivations in a section marked "Proofs". The status of these arguments is that they deserve more attention than a technical appendix, but at the same time they are not entirely necessary for expositional continuity. Therefore a reader uninterested in the deeper arguments can always skip this section and read on, relatively unscathed.

I turn now to the proofs.

We will need to fix threshold values of the discount factor, and then settle on the maximum of these. The lemmas that follow describe the different thresholds. To this end, define, for any substructure $\mathbf{n}$ and δ,

$$r(\mathbf{n}, \delta) = \delta v(1, c(\mathbf{n}.1)) \qquad \text{if } t(\mathbf{n}) = 1$$

(5.6)
$$= a(\mathbf{n}) \frac{\delta \rho(s) t(\mathbf{n})}{(1 - \delta) + \delta \rho(s) t(\mathbf{n})} \qquad \text{if } t(\mathbf{n}) \geq 2,$$

where s, as before, is just $n - K(\mathbf{n})$, and $a(\mathbf{n})$ is the average worth generated at "stage $\mathbf{n}$" in the algorithm (see (5.2)).

Though this comment is unnecessary for the formal developments below, the numbers $r(\mathbf{n}, \delta)$ will serve as standard-equilibrium rejection thresholds for agents receiving offers after the substructure $\mathbf{n}$ has left. We begin with

LEMMA 5.1. *There exists $\delta^1 \in (0, 1)$ such that for any $\delta \in (\delta^1, 1)$, and any substructure $\mathbf{n}$, $t(\mathbf{n})$ uniquely maximizes (in t) the expression*

(5.7)
$$\frac{v(t, c(\mathbf{n}.t))}{(1 - \delta) + \delta \rho(s) t}$$

where $s = n - K(\mathbf{n})$, and also the expression

(5.8) $$v(t, c(\mathbf{n}.t)) - (t-1)r(\mathbf{n}, \delta).$$

Proof. Fix $\mathbf{n}$ and consider the maximization of the expression in (5.7). By the well-known maximum theorem, the set of maximizers is upper hemicontinuous in δ, and so all limit points of sequences from this set must lie in the set of maximizers of

$$\frac{v(t, c(\mathbf{n}.t))}{t},$$

as $\delta \to 1$. But the maximizers are just integers from some finite set, so in fact the collection of maximizers of (5.7) — call it $\mu(\mathbf{n}, \delta)$ — must become a *subset* of $\mu(\mathbf{n}, 1)$ for all δ large enough; bigger than some $\delta^1 \in (0, 1)$, where δ^1 can be chosen uniformly over the finitely many different substructures.

So for every $\delta > \delta^1$ and for every maximizer t of (5.7) the value of

$$\frac{v(t, c(\mathbf{n}.t))}{t}$$

is just the same (and at its maximum). Comparing this expression and (5.7), it is obvious, then, that the maximizer of (5.7) must indeed exhibit the largest value of

$$\frac{t}{(1 - \delta) + \delta\rho(s)t}$$

among all values of t that achieve the maximum of $v(t, c(\mathbf{n}.t))/t$. But this fraction strictly increases in t, so we've proved that for all $\delta > \delta^1$ there is only one maximizer of (5.7), and that is $t(\mathbf{n})$.

To establish the maximization of (5.8), notice that we can scale up and rewrite (5.7) as

$$\frac{[(1 - \delta) + \delta\rho(s)]v(t, c(\mathbf{n}.t))}{(1 - \delta) + \delta\rho(s)t} = v(t, c(\mathbf{n}.t)) - (t - 1)\frac{\delta\rho(s)v(t, c(\mathbf{n}.t))}{(1 - \delta) + \delta\rho(s)t}$$

(5.9) $$\equiv v(t, c(\mathbf{n}.t)) - (t - 1)y(t),$$

where

$$y(t) \equiv \frac{\delta\rho(s)v(t, c(\mathbf{n}.t))}{(1 - \delta) + \delta\rho(s)t}.$$

Because $t(\mathbf{n})$ maximizes (5.7), it is immediate that

$$y(t) = \frac{\delta\rho(s)v(t, c(\mathbf{n}.t))}{(1 - \delta) + \delta\rho(s)t} \leq \frac{\delta\rho(s)v(t(\mathbf{n}), c(\mathbf{n}.t(\mathbf{n})))}{(1 - \delta) + \delta\rho(s)t(\mathbf{n})} = \frac{\delta\rho(s)t(\mathbf{n})a(\mathbf{n})}{(1 - \delta) + \delta\rho(s)t(\mathbf{n})},$$

and this last expression equals $r(\mathbf{n}, \delta)$ when $t(\mathbf{n}) \geq 2$, while it is bounded above by $r(\mathbf{n}, \delta)$ when $t(\mathbf{n}) = 1$ (examine (5.6)). So $y(t) \leq r(\mathbf{n}, \delta)$ in all cases. Using this information in (5.9) and remembering again that $t(\mathbf{n})$ uniquely maximizes (5.7), we may conclude that

$$v(t(\mathbf{n}), c(\mathbf{n}.t(\mathbf{n}))) - [t(\mathbf{n}) - 1]r(\mathbf{n}, \delta) > v(t, c(\mathbf{n}.t)) - (t-1)y(t)$$
$$\geq v(t, c(\mathbf{n}.t)) - (t-1)r(\mathbf{n}, \delta),$$

which establishes the unique maximization of (5.8) at $t = t(\mathbf{n})$.   □

LEMMA 5.2. *Assume that the protocol is rejector-friendly. Then there exists* $\delta^2 \in (0, 1)$ *such that for all* $\delta \in (\delta^2, 1)$,

$$(5.10) \qquad\qquad r(\mathbf{n}, \delta) > \delta v(1, c(\mathbf{n}.1))$$

whenever $t(\mathbf{n}) \geq 2$.

Proof. Recall that for $t(\mathbf{n}) \geq 2$,

$$r(\mathbf{n}, \delta) = a(\mathbf{n})\frac{\delta \rho(s)t(\mathbf{n})}{(1 - \delta) + \delta \rho(s)t(\mathbf{n})},$$

so that (5.10) will hold whenever

$$(5.11) \qquad\qquad a(\mathbf{n})\frac{\delta \psi}{(1 - \delta) + \delta \psi} > \delta v(1, c(\mathbf{n}.1)),$$

where $\psi \equiv \rho(s)t(\mathbf{n})$.

Because $t(\mathbf{n}) \geq 2$ by assumption, $a(\mathbf{n})$ — the maximal value of the ratio $v(t, c(\mathbf{n}.t))/t$ — is no less than $v(1, c(\mathbf{n}.1))$. If strict inequality holds, it is obvious that there is a threshold $\delta(\mathbf{n}) \in (0, 1)$ such that (5.11) — and therefore (5.10) — holds for all $\delta \geq \delta(\mathbf{n})$.

If equality holds — i.e., $a(\mathbf{n}) = v(1, c(\mathbf{n}.1)) = z$, say — then a similar threshold can be obtained but the argument is more subtle. Differentiate the left and right hand sides of (5.11) with respect to δ and evaluate these two derivatives at the limit $\delta = 1$. The limit derivative on the left is z/ψ, while the one on the right is just z. Now observe that $\psi = \rho(s)t(\mathbf{n}) > 1$, because $\rho(s) > 1/2$ by rejector-friendliness and $t(\mathbf{n}) \geq 2$ by assumption. It follows once again that there is $\delta(\mathbf{n}) \in (0, 1)$ such that (5.10) holds for all $\delta \geq \delta(\mathbf{n})$.

To complete the proof, note that because there are finitely many $\mathbf{n}$, the threshold can be made uniform by selecting $\delta^2 = \max_{\mathbf{n}} \delta(\mathbf{n})$.   □

Remark. Rejector-friendliness can be dispensed with in this lemma if $a(\mathbf{n}) > v(1, c(\mathbf{n}.1))$ whenever $t(\mathbf{n}) \geq 2$.

To continue, define

(5.12) $$a(\mathbf{n}, \delta) \equiv a(\mathbf{n}) \frac{[(1 - \delta) + \delta \rho(s)] t(\mathbf{n})}{(1 - \delta) + \delta \rho(s) t(\mathbf{n})},$$

where, as before, $s = n - K(\mathbf{n})$. As an informal comment, $a(\mathbf{n}, \delta)$ will represent the equilibrium payoff to a proposer under a standard equilibrium (compare with the definition of $r(\mathbf{n}, \delta)$, the equilibrium payoff to a *responder*, in (5.6)).

LEMMA 5.3. *For all δ and substructures $\mathbf{n}$, $\delta a(\mathbf{n}, \delta) \geq r(\mathbf{n}, \delta)$.*

Proof. If $t(\mathbf{n}) = 1$ this is trivially true because $a(\mathbf{n}, \delta) = v(1, c(\mathbf{n}.1))$ and $r(\mathbf{n}, \delta) = \delta v(1, c(\mathbf{n}.1))$. The case $t(\mathbf{n}) \geq 2$ is also simple: compare (5.6) and (5.12) and note that $\delta[(1 - \delta) + \delta \rho(s)] \geq \delta \rho(s)$. □

LEMMA 5.4. *For any substructure $\mathbf{n}$ and nonnegative integer M,*

(5.13) $$\lim_{\delta \to 1} \frac{d \delta^M a(\mathbf{n}, \delta)}{d \delta} = a(\mathbf{n}) \left[M - \frac{t(\mathbf{n}) - 1}{\rho(s) t(\mathbf{n})} \right],$$

and

(5.14) $$\lim_{\delta \to 1} \frac{d \delta r(\mathbf{n}, \delta)}{d \delta} = a(\mathbf{n}) \left[1 + \frac{1}{\rho(s) t(\mathbf{n})} \right],$$

where $s = n - K(\mathbf{n})$.

Proof. (5.13) and (5.14) follow from elementary differentiation with respect to δ in (5.12) and (5.6) respectively; I omit the details. □

Say that a substructure $\mathbf{n}'$ is a *continuation* of another substructure $\mathbf{n}'$ if $\mathbf{n}'$ can be reached from $\mathbf{n}$ by repeatedly applying the algorithm from $\mathbf{n}$.

LEMMA 5.5. *Assume NAW and a rejector-friendly protocol. Then there exists $\delta^3 \in (0, 1)$ such that for all $\delta \in (\delta^3, 1)$, for every pair of substructures $(\mathbf{n}, \mathbf{n}')$ with $\mathbf{n}'$ a continuation of $\mathbf{n}$, and for every weight $\mu \in [0, 1]$ with $\mu \leq 1/[t(\mathbf{n}') - 1]$,*

(5.15) $$a(\mathbf{n}, \delta) > \delta \left[\mu a(\mathbf{n}', \delta) + (1 - \mu) r(\mathbf{n}', \delta) \right].$$

Proof. By NAW, $a(\mathbf{n}) \geq a(\mathbf{n}')$. If strict inequality holds, we can easily find a discount threshold $\delta(\mathbf{n}, \mathbf{n}')$ such that (5.15) holds whenever $\delta > \delta(\mathbf{n}, \mathbf{n}')$.[7] So consider the case in which $a(\mathbf{n}) = a(\mathbf{n}') = a$, say.

---

[7] After all, both $a(\mathbf{n}'', \delta)$ and $r(\mathbf{n}'', \delta)$ converge to $a(\mathbf{n}'')$ as $\delta \to 1$ for every substructure $\mathbf{n}''$.

Write $t' = t(\mathbf{n}')$ to ease the notation. Applying Lemma 5.4 and noting that $s' = n - K(\mathbf{n}')$, we see that

$$\lim_{\delta \to 1} \frac{d\delta\left[\mu a(\mathbf{n}', \delta) + (1 - \mu)r(\mathbf{n}', \delta)\right]}{d\delta} = a - \mu \frac{a(t' - 1)}{\rho(s')t'} + (1 - \mu)\frac{a}{\rho(s')t'}.$$

This is obviously strictly positive whenever $t' = 1$. If $t' \geq 2$,

$$\lim_{\delta \to 1} \frac{d\delta\left[\mu a(\mathbf{n}', \delta) + (1 - \mu)r(\mathbf{n}', \delta)\right]}{d\delta} = a - \mu \frac{a(t' - 1)}{\rho(s')t'} + (1 - \mu)\frac{a}{\rho(s')t'}$$

$$\geq a - \frac{a(t' - 1)}{(t' - 1)\rho(s')t'} + \frac{t' - 2}{t' - 1}\frac{a}{\rho(s')t'}$$

(5.16) $$= a\left[1 - \frac{1}{t'(t' - 1)\rho(s')}\right] > 0$$

where the very last inequality follows from the fact that $t' \geq 2$ and $\rho(s') > 1/2$. So the above limit is strictly positive for *all* values of t'.

At the same time, Lemma 5.4 also tells us that

(5.17) $$\lim_{\delta \to 1} \frac{da(\mathbf{n}, \delta)}{d\delta} = -a\frac{t(\mathbf{n}) - 1}{\rho(s)t(\mathbf{n})} \leq 0.$$

The two opposing inequalities (5.16) and (5.17) imply that there exists $\delta(\mathbf{n}, \mathbf{n}')$ such that (5.15) holds for all $\delta > \delta(\mathbf{n}, \mathbf{n}')$. To complete the proof, take $\delta^3 = \max \delta(\mathbf{n}, \mathbf{n}')$ over all $(\mathbf{n}, \mathbf{n}')$ satisfying the conditions of the lemma. $\qquad\square$

Remark. Note that if NAW holds with strict inequality, rejector-friendliness can be dispensed with in this lemma.

Proof of Proposition 5.1. In what follows, we take the threshold δ^* to be the maximum of the three values δ^1, δ^2 and δ^3 identified in Lemmas 5.1, 5.2, and 5.5. As required, we assume NAW, and prove that a standard equilibrium exists for all rejector-friendly protocols whenever $\delta \geq \delta^*$.

We describe a strategy profile. In what follows, index every individual by a number i, $i = 1, \ldots, n$. *Counting from i* will mean looking at progressively higher indices from i, modulo n. (So, for instance, n and then 1 are the two closest individuals counting from $n - 1$.)

Suppose that a partial coalition structure π has left; let $\mathbf{n}$ be the numerical structure corresponding to it.

I describe proposer actions. The proposer chooses a coalition of size $t = t(\mathbf{n})$, the integer identified in the algorithm. The exact choice of coalition will be described below (she must include herself of course). She makes a proposal to this coalition which gives each member (other than herself) the value $r(\mathbf{n}, \delta)$, and gives herself the value $v(t, c(\mathbf{n}.t)) - (t - 1)r(\mathbf{n}, \delta)$, which is easily checked, using (5.6) and (5.12), to be $a(\mathbf{n}, \delta)$.[8]

If $t = 1$, then there is no choice of coalition involved; the proposer effectively exits as a one-person coalition. If $t > 1$ and the proposer is the *first* individual to make a proposal following the exit of $\mathbf{n}$, she includes herself in the coalition and the $t-1$ closest active individuals to her, counting from herself. Finally, if $t > 1$ and a previous proposal following the exit of $\mathbf{n}$ has been rejected, the proposer includes herself, the rejector of the *latest* proposal, and the $t - 2$ other closest active individuals to her, counting from herself.

I now turn to responses at stage $\mathbf{n}$. Recall the threshold $r(\mathbf{n}, \delta)$ defined in (5.6). *Provided that all subsequent respondents to the same proposal are offered at least $r(\mathbf{n}, \delta)$ each,* a responder accepts the proposal *if and only if* it offers her at least $r(\mathbf{n}, \delta)$.

This specification guarantees that a proposal is accepted if and only if every responder is given at least $r(\mathbf{n}, \delta)$. For completeness, I will still need to specify what happens if a subsequent responder is offered less. Before I do so, I describe a rejector's unilateral exit actions.

A rejector can unilaterally exit as a one-person coalition or stay on (with a proposer subsequently chosen according to the given protocol). We specify that she must stay on if $t(\mathbf{n}) \geq 2$. If $t(\mathbf{n}) = 1$, she computes the two payoffs from unilateral exit and staying on, which are fully defined given our specification so far. She chooses the option with the higher payoff, breaking indifference by unilateral exit.

Finally, to specify the identity of a rejector, we need to consider the one remaining case. Consider a responder to a proposal in which some *subsequent* responder is offered strictly less than $r(\mathbf{n}, \delta)$. Given our prescribed actions so far, such a proposal *will* be rejected. Whether or not our responder should do the rejecting is a case of

---

[8]To be precise, remembering that a proposal is really a contingent plan, our propose offers to each partner a payoff of $r(\mathbf{n}, \delta)$ in the event that the numerical coalition structure $c(\mathbf{n})$ is formed, and any other payoff division otherwise.

tedious but elementary computation. If she rejects, her present-value payoff is $r(\mathbf{n}, \delta)$ by construction. If she accepts, an inductive step tells us the identity of the later rejector, and allows us to compute subsequent present-value payoffs for our responder. We require that she accept the proposal if this latter payoff is greater, and reject it otherwise.[9]

We have fully described a strategy profile. We now show that it is an equilibrium. We will begin with the proposer at some stage $\mathbf{n}$; call her P. Because everyone is using exactly the same accept–reject threshold, P will not care whom she includes, so all we need to check is that the prescribed coalition *size* is optimal, given the rest of the strategy profile. If the P makes an acceptable proposal, the optimality of $t(\mathbf{n})$ follows right away from the second part of Lemma 5.1. Otherwise, she makes an unacceptable proposal. Applying the equilibrium strategy profile from that point on we note the following consequences:

(a) a delay discounted using the factor δ,

(b) no further delays,

(c) at each future stage (indexed by $k = 1, 2, \ldots$), conditional on not having already exited, P is chosen to be proposer again with probability f_k, is not chosen as proposer but nevertheless included in a proposal with probability g_k, and passed on to the next stage with remaining probability.

Therefore, *conditional* on P being included at some future stage k, the present-value expected return to her is

(5.18) $\delta\left[\mu_k a(\mathbf{n}_k, \delta) + (1 - \mu_k) r(\mathbf{n}_k, \delta)\right],$

---

[9]A bit more formally, proceed inductively as follows. Certainly the last responder in the response ordering to be offered less than $r(\mathbf{n}, \delta)$ will reject the proposal if it ever gets to her for approval. Inductively, pick a particular responder R and suppose that all responders following R in the order have been assigned accept-reject decisions for the proposal in question. If R rejects, our prescribed strategy yields her an expected value of $r(\mathbf{n}, \delta)$. If she accepts, an expected value is also well-defined, given the inductive step and our specification of the proposal. I should add that in most cases, it will be in R's interest to grab the initiative and do the rejecting herself. But given the generality of our model, it is possible that in some situations that she will benefit from passing the buck to another rejector.

where n_k is the numerical coalition structure at the start of that stage, and $\mu_k \equiv f_k/(f_k + g_k)$ is the appropriate conditional probability that P will be proposer.

I claim that for each such stage with $t_k \equiv t(\mathbf{n}_k)$, $\mu_k \leq 1/(t_k - 1)$.

To see this, assume $t_k \geq 2$ (there is nothing to prove if $t_k = 1$). First consider any stage $k \geq 2$, that is, a stage *after* the stage immediately following the unacceptable proposal. Let s_k be the number of active players. Then $f_k = 1/s_k$ (P is chosen equiprobably to be proposer again). What are her chances of being chosen to be a responder instead? There are clearly $t_k - 1$ other individuals such that P is among the closest active $t_k - 1$ from each of them: given our prescribed strategy profile, these and exactly these individuals will include P as responder. So $g_k = (t_k - 1)/s_k$, and therefore $\mu_k = 1/t_k \leq 1/(t_k - 1)$, as claimed.

Now consider the stage immediately following the unacceptable proposal. Obviously $s_1 \geq 2$. Let R be the rejector of the unacceptable proposal. Note that P is chosen again to propose with probability $f_1 = [1 - \rho(s_1)]/(s_1 - 1)$ (because R is *not* chosen as proposer with probability $1 - \rho(s_1)$, and this residual is distributed equally among the other active agents). Now we need to estimate g_1, the probability that P will be called upon to respond. Notice that there are $t_1 - 2$ other individuals such that P is among the closest active $t_1 - 2$ from each of them; these will surely include P. The probability of this happening is *at least* $[1 - \rho(s_1)](t_1 - 2)/(s_1 - 1)$, and it will be even more if R is among these $t_1 - 2$ individuals. Therefore $g_1 \geq [1 - \rho(s_1)](t_1 - 2)/(s_1 - 1)$, and so $\mu_1 \leq 1/(t_1 - 1)$, completing the proof of the claim.

Combine the claim, expression (5.18) and Lemma 5.5. It follows immediately that P *cannot* gain by making an unacceptable proposal.

Now we turn to responses (and associated unilateral exit decisions). Suppose that the substructure $\mathbf{n}$ has left, and a proposal is on the table. First suppose that all subsequent responders are offered at least $r(\mathbf{n}, \delta)$. Then from the construction of $r(\mathbf{n}, \delta)$ (see (5.6)) as well as the presumption that the specified strategy profile will be followed subsequently, it must be optimal for our respondent to accept if she is offered at least $r(\mathbf{n}, \delta)$, and to reject otherwise. The remaining case, in which every subsequent responder *is* not offered at least $r(\mathbf{n}, \delta)$ is optimally specified by construction, assuming that the going strategy profile is followed thereafter.

Finally, consider exit decisions. Suppose that $t(\mathbf{n}) \geq 2$. Then the optimality of staying on (as opposed to exit) is guaranteed by Lemma 5.2. If $t(\mathbf{n}) = 1$, optimality is guaranteed by construction.

We have therefore established that the prescribed strategy profile is indeed an equilibrium for all $\delta \in (\delta^*, 1)$. It is clearly standard, for it involves pure strategies and at every stage the proposer makes an acceptable offer. □

Remark. By the remarks following Lemmas 5.2 and 5.5, the rejector-friendliness restriction on protocols can be dropped if NAW holds strictly at every stage. All we need is that the rejector gets to propose with positive probability following a rejection.

We now prepare for the proof of Proposition 5.2. Note that under the rejector-proposes protocol (RPP), and at every stage at which a coalition structure π has already formed, an equilibrium payoff is well-defined for every active player in her role as proposer at this stage. Let x_j denote this particular payoff for each player j.

LEMMA 5.6. *Assume RPP, the rejector-proposes protocol. Consider an equilibrium, and a stage in which the structure π has left the game, with associated numerical structure $\mathbf{n}$. Suppose further that if a coalition of size t forms, the equilibrium numerical structure following $\mathbf{n}.t$ is $c(\mathbf{n}.t)$.*

Then if player i as proposer makes an acceptable proposal with positive probability in equilibrium, $x_k \geq x_i$ for every active player k.

Proof. Let i make an acceptable proposal to T with positive probability. Pick active $k \neq i$. If $k \notin T$, k can make an offer to $\{T - i\} \cup k$, and give everyone slightly more than what i was giving them; this is *strictly* acceptable to all (for post-acceptance the continuation numerical structure is exactly the same by assumption). Therefore $x_k \geq x_i - \epsilon$ for all $\epsilon > 0$, or $x_k \geq x_i$.

On the other hand, if $k \in T$, k can acceptably propose T (again, an ϵ-argument of the sort above will suffice). Consequently, writing t for the cardinality of T,

$$x_k \geq v(t, c(\mathbf{n}.t)) - \delta \sum_{j \in T; j \neq k} x_j$$

$$= v(t, c(\mathbf{n}.t)) - \delta \sum_{j \in T; j \neq i} x_j + \delta x_k - \delta x_i$$

$$= x_i + \delta x_k - \delta x_i,$$

where the first line uses RPP (active players must be paid off their equilibrium payoff as proposer), and the last line follows from the fact that i's proposal to t does attain her equilibrium payoff. Now rearrange this inequality to see that

$$x_k - x_i \geq \delta(x_k - x_i),$$

which simply means that $x_k \geq x_i$. □

Proof of Proposition 5.2. Consider the RPP. Fix a standard equilibrium, and let $\delta \in (\delta^1, 1)$, where δ^1 is set in the proof of Lemma 5.1. We proceed by induction on the cardinality of the set of active players. If there is one active player left, then there is nothing to prove. Inductively, suppose that for every stage $\mathbf{n}$ with $K(\mathbf{n}(\pi)) = m + 1, \ldots, n - 1$ for some $m \geq 0$, the equilibrium coalition structure generated is $c(\mathbf{n})$. Consider a stage π with associated numerical structure $\mathbf{n}$, such that with $K(\mathbf{n}) = m$. Let T^* (with cardinality t^*) be a coalition that forms at this stage. We need to prove that

(5.19) $$t^* = t(\mathbf{n}).$$

Because the equilibrium is standard, *every* active player makes an acceptable proposal with positive probability. By the induction hypothesis and Lemma 5.6, $x_j = x_i \equiv x$ for all active i and j. Therefore, invoking induction again as well as the optimality of the proposal leading to T^*,

(5.20) $$x = v(t^*, c(\mathbf{n}.t^*)) - \delta(t^* - 1)x \geq v(t, c(\mathbf{n}.t)) - \delta(t - 1)x,$$

for all $t \in \{1, \ldots, n - K(\mathbf{n})\}$.

I claim now that t^* maximizes the expression in (5.7) with $\rho(s) = 1$ (by RPP); that is, the ratio

$$\frac{v(t, c(\mathbf{n}.t))}{1 + \delta(t - 1)}.$$

Suppose not; then there is t such that

$$\frac{v(t, c(\mathbf{n}.t))}{1 + \delta(t - 1)} > \frac{v(t^*, c(\mathbf{n}.t^*))}{1 + \delta(t^* - 1)} = x,$$

so that — rearranging terms —

$$v(t, c(\mathbf{n}.t) - \delta(t - 1)x > x,$$

but this contradicts (5.20). So t^* maximizes the expression in (5.7) but then we are done, by the first part of Lemma 5.1. □

A much easier variant of Lemma 5.5 applies to the RPP when SNAW holds. Say that a substructure $\mathbf{n}'$ is a *weak continuation* of another substructure $\mathbf{n}'$ if $\mathbf{n}'$ can be reached from $\mathbf{n}$ by first concatenating any restricted maximizer t^* at $\mathbf{n}$ to obtain $\mathbf{n}.t^*$, and thereafter applying the algorithm.

LEMMA 5.7. *Assume SNAW and the RPP. Then there exists $\delta^4 \in (0, 1)$ such that for all $\delta \in (\delta^4, 1)$, for every pair of substructures $(\mathbf{n}, \mathbf{n}')$ with $\mathbf{n}'$ a weak continuation of $\mathbf{n}$,*

(5.21) $$a(\mathbf{n}, \delta) > \delta a(\mathbf{n}', \delta).$$

Proof. Because $\lim_{\delta \to 1} a(\mathbf{n}'', \delta) = a(\mathbf{n}'')$ for any substructure $\mathbf{n}''$, the case $a(\mathbf{n}) > a(\mathbf{n}')$ is trivial to handle. By SNAW, the only other possibility is $a(\mathbf{n}) = a(\mathbf{n}')$. In this case, apply (5.13) to $a(\mathbf{n}, \delta)$ for $M = 0$ and then again to $a(\mathbf{n}', \delta)$ for $M = 1$. It is immediate that

$$\lim_{\delta \to 1} a(\mathbf{n}, \delta) > \lim_{\delta \to 1} \delta a(\mathbf{n}', \delta),$$

so that there exists a threshold $\delta(\mathbf{n}, \mathbf{n}')$ such that (5.21) holds for all $\delta > \delta(\mathbf{n}, \mathbf{n}')$. Pick $\delta^4 = \max \delta(\mathbf{n}, \mathbf{n}')$ over all the relevant pairs to complete the proof. ☐

Proof of Proposition 5.4. First we fix a threshold δ. Pick $\delta^5 \in (0, 1)$ such that for all $\delta \in (\delta^5, 1)$, $a(\mathbf{n}, \delta) > a(\mathbf{n}', \delta)$ whenever $a(\mathbf{n}) > a(\mathbf{n}')$. Take $\hat{\delta} \equiv \max\{\delta^4, \delta^5\}$, where we've defined δ^4 in Lemma 5.7.

Fix any equilibrium. We will show that it must be standard. The proof is by induction on the cardinality of the set of active players. At every stage when there is only one active player left, the subgame equilibrium is trivially standard. Now suppose that whenever there are fewer than s active players, the subgame equilibrium is standard. Consider any stage with s active players. Let x_i be the equilibrium payoff to active player i, if she is the proposer at this stage. Name the players such that $x_1 \leq \ldots \leq x_s$.

Now, *some* player must make an acceptable offer at this stage.[10] By Lemma 5.6 we can take this player to be player 1. We can also

---

[10]Recall our normalization that all coalitional worths are nonnegative, and that eternal disagreement yields negative payoffs.

without any loss of generality take the acceptable offer to be made to a coalition of the form $T^* = \{1, \dots, t^*\}$.[11]

We claim that

(5.22) $$\frac{v(t, c(\mathbf{n}.t))}{t} \leq \frac{v(t^*, c(\mathbf{n}.t^*))}{t^*} \text{ for all } t \leq t^*.$$

To prove this claim, note first that by virtue of player 1 making an acceptable equilibrium proposal to T^*,

(5.23) $x_1 = v(t^*, c(\mathbf{n}.t^*)) - \delta \sum_{j=2}^{t^*} x_j \geq v(t, c(\mathbf{n}.t)) - \delta \sum_{j=2}^{t} x_j$ for all $j \in S$.

for all $t \leq s$.

Suppose, now, that (5.22) is false. Then there exists $\hat{t} < t^*$ such that $\frac{v(\hat{t}, c(\mathbf{n}.\hat{t}))}{\hat{t}} > \frac{v(t^*, c(\mathbf{n}.t^*))}{t^*}$. Because $\delta > \delta^4$,

$$\hat{a} \equiv \frac{v(\hat{t}, c(\mathbf{n}.\hat{t})}{1 + \delta(\hat{t} - 1)} > a^* \equiv \frac{v(t^*, c(\mathbf{n}.t^*))}{1 + \delta(t^* - 1)},$$

or

$$v(\hat{t}, c(\mathbf{n}.\hat{t})) - \delta(\hat{t} - 1)\hat{a} > v(t^*, c(\mathbf{n}.t^*)) - \delta(t^* - 1)a^*.$$

Since $\hat{a} > a^*$, we may combine this inequality with (5.23) and rearrange terms to see that

$$(t^* - \hat{t})\hat{a} > \sum_{j=\hat{t}+1}^{t^*} x_j,$$

which permits us to conclude that

(5.24) $$\hat{a} > x_j \text{ for all } j = 1, \dots, \hat{t}.$$

However, (5.24) implies that

$$v(\hat{t}, c(\mathbf{n}.\hat{t})) - \delta \sum_{j=2}^{\hat{t}} x_j > v(\hat{t}, c(\mathbf{n}.\hat{t})) - \delta(\hat{t} - 1)\hat{a} = \hat{a} > x_1,$$

which contradicts (5.23) and so establishes (5.22).

An immediate consequence of the claim is that t^* must be a restricted maximizer.

---

[11]The reason is simple. Under RPP, and given the induction hypothesis, if a proposer i includes some $j \neq i$ in an acceptable coalition, she *must* include all k such that $x_k < x_j$.

Suppose, now, that the theorem is false, i.e., there exists a player who makes an unacceptable offer. By Lemma 5.6 once again and our ordering of players, there is no loss of generality in assuming that this is player s. There is, moreover, no sense in making that offer to a player who will make another unacceptable offer for sure, so we may as well suppose that player 1 is the first rejector of this offer. Notice also that $s \notin T^*$, the coalition chosen by player 1.[12] By the induction hypothesis, the algorithm applies thereafter, and s is absorbed later at some weak continuation $\mathbf{n'}$ of $\mathbf{n}$.

Applying the induction hypothesis again,

(5.25) $x_s \le \delta a(\mathbf{n}.t^* \ldots t_{k-1}, \delta).$

Moreover, by SNAW and (5.22), which established t^* as a restricted maximizer, $a(\mathbf{n'}) \le a(\mathbf{n})$. Applying Lemma 5.7, we conclude that for all $\delta > \hat{\delta}$,

(5.26) $x_s \le \delta a(\mathbf{n}.t^* \ldots t_{k-1}, \delta) < a(\mathbf{n}, \delta) \equiv a.$

Since $a + \delta(t(\mathbf{n}) - 1)a = v(t(\mathbf{n}), c(\mathbf{n}))$, (5.26) implies that

$$v(t(\mathbf{n}), c(\mathbf{n})) - \delta \sum_{j=2}^{t(\mathbf{n})} x_j > x_1,$$

which contradicts (5.23).   □

5.6 Summary

In this chapter, we've studied the special case of symmetric coalitional bargaining games with transferable payoffs. The main restriction is really symmetry; under weak conditions, the main results carry over to the NTU case.

We begin this chapter by introducing an algorithm which works for *all* symmetric TU partition functions. The algorithm is formally described in Section 5.2, and generates a distinguished coalition structure by recursively maximizing average coalitional worth. The recursive definition is necessary because the "average worth" of a coalition depends on the ambient structure within which that coalition is embedded. By supposing that average worth maximization

---

[12]Otherwise player s's expected payoff would be precisely δx_s rather than x_s.

will be carried out in all later stages as well, we can "predict" that ambient structure.

The main task of the chapter is to connect the coalition structure of this algorithm to the actual equilibria of coalitional bargaining. To this end, we identify a particular kind of equilibrium, which we call *standard equilibrium*. In this equilibrium, every proposer makes an acceptable proposal with positive probability at every stage. Under the restriction that the algorithmically generated average worth is weakly decreasing from one stage of the algorithm to the next ("condition NAW"), a standard equilibrium exists. What is more, over a wide variety of protocols, there is a standard equilibrium that yields the distinguished coalition structure identified in the algorithm (Proposition 5.1).

Two more results tighten the characterization further. First, the algorithmically identified coalition structure is not only common to standard equilibrium regardless of the particular protocol, it is also the unique such structure (Proposition 5.2). There isn't any other structure which can be generated by standard equilibria over the same range of protocols. Second, under an additional condition *no other equilibrium exists*, standard or otherwise, in the rejector-proposes protocol (Proposition 5.4). The additional condition is a stronger version of the NAW property.

I claim that the conditions required to establish these results are acceptably mild, and satisfied in a variety of contexts. But the proof of the pudding lies in the eating, and so we turn to two applications in the next chapter.

CHAPTER 6

Applications

The analysis of the previous chapter may be viewed as cooperative game theory redone for games with externalities (though we have restricted our attention so far to symmetric games). This opens the door to a variety of applications, something that the use of characteristic functions cannot achieve. In this chapter we study two such applications.

6.1 Cournot Oligopoly

This section takes up a textbook example of symmetric Cournot oligopoly.[1] The difference from the textbook exposition is that we study binding agreements, as opposed to noncooperative equilibria.

6.1.1 Model. Suppose that n oligopolists produce a quantity x of a homogeneous product, the price P of which is determined by a linear demand curve: $P = A - bx$. Assume that there is a fixed unit cost c of production, common to all firms.

In Chapter 3 we've already derived the partition function for this game. The idea is simple: if the firms group themselves into m distinct coalitions or cartels, with each cartel agreeing to write a binding agreement, then each such coalition will behave like an independent firm. So the worth of each cartel is just the equilibrium firm payoff in the m-firm Cournot oligopoly. This yields a symmetric TU partition function: for each cartel or coalition S and coalition

---

[1]The analysis of this section is based on Bloch (1996) and Ray and Vohra (1999).

structure π with $S \in \pi$,

$$(6.1) \qquad v(S, \pi) = \frac{1}{(m(\pi) + 1)^2},$$

where $m(\pi)$ is the number of coalitions in π, and where the parameters A, b and c have been normalized so that $\frac{(A-c)^2}{b} = 1$.

6.1.2 Intuitive Approach. Chapter 3, Section 2.5.1, discusses the three-firm version of this game, arguing that a single firm might want to break off negotiations to form a cartel, especially if it's under the illusion that the other two firms will agree to write a binding agreement. But that's what illusions are: illusory, for the remaining two firms will need to share the duopoly payoff, which — by (6.1) — is 1/9. At least *one* of them is going to have to settle for no more than 1/18, but that means that the two remaining firms are better off splitting and picking up the "triopoly" payoff of 1/16 each. With this in mind, no firm should attempt to break off negotiations to begin with, and the grand cartel indeed forms.

This intuitive approach goes well with the algorithm. It is easy enough to check that when $n = 3$, the algorithmic structure is given by $\mathbf{n}^* = \{3\}$.

However, in taking this intuitive argument further, the various possibilities multiply. All sorts of other deviations need to be entertained, along with predictions about what will happen following those deviations. Remember, apart from the assumption that equilibria are stationary Markovian, we are allowing for all kinds of strategy profiles, including those that are asymmetric and those that require mixing. This is where the intuitive approach runs into limitations, and the results of Chapter 5 play a productive role.

6.1.3 Using the Algorithm. The first item of business is to verify that algorithmic average worths satisfy the criteria described in Chapter 5, and indeed they do: SNAW holds, as we shall see formally below. We therefore have the following proposition, which is partly a consequence of SNAW and also describes the equilibrium coalition structure.

PROPOSITION 6.1. *The only equilibrium coalition structure common to all rejector-friendly protocols (as bargaining frictions vanish) is the algorithmically derived structure* $\mathbf{n}^*$. *It consists of L singleton firms and a single cartel of size* $n - L$, *where L is the smallest nonnegative integer such*

that
$$n - L < (L + 2)^2 + 1.$$

In particular, there is full cartelization in this example whenever there are four firms or less, and imperfect cartelization thereafter.

For some intuition, we invoke an important observation due originally to Salant, Switzer and Reynolds (1983): If several firms are already out of a potential cartel, and the number of firms left is "small enough", then the remaining firms will not find it advantageous to form a cartel. Intuitively, the gain in market concentration does not justify the profit-sharing that will be needed. Applying this idea recursively to the remaining number of players, we can find a threshold at which the average payoff to the remaining players, if they stay together, is approximately the same as when a player quits, sparking off a cartel collapse.

Summarizing so far, we see that at this threshold, firms would rather stay together than break up. But *knowing this is so*, those firms in excess of this threshold will not agree to join the cartel, predicting — correctly — that the remaining firms will stay together. This creates an equilibrium outcome with one large cartel and several singleton firms.

We relegate a formal proof of this proposition to the end of the chapter.

6.1.4 On Efficiency. This result hints at a feature that we will explore in greater detail in the next example. Note that full cartelization — the formation of the grand coalition of all firms — is efficient as far as the firms are concerned.[2] Yet the outcome isn't efficient for $n \geq 4$, despite the fact that all agents are free to make any proposal they wish to anyone else. What creates the inefficiency is the fact that *the very act of making a proposal opens the door to possible counteroffers*. These potential counteroffers must be satisfactorily dealt with in the proposal: the proposer must give away part of the social surplus when a group is formed. This drives a wedge between the proposer's incentives and the socially efficient outcome.

Several remarks are in order. First, the wedge is not inevitable, or at least it doesn't have to be sharp enough so that it inevitably causes

---

[2]This is not the case for consumers, but that isn't the point. The firms make up the set of relevant agents for this example.

an efficiency failure. (Recall that the grand coalition *does* emerge when there are four firms or less.)

Second, the particular form that the wedge assumes in our theory is sufficient for our purposes, but it is not necessary. To explain this further, observe that in the standard equilibrium common to all protocols, the proposer is so wary of possible counteroffers that coalitions that do form are driven to divide their payoff equally. This is why average worths play an important role in our propositions. One might imagine a world, however, in which the fear of counteroffers isn't that stark, in which *some* sidepayments have to be made but not to the extent of equal division. (We will see more of this in the general asymmetric model.) But the potential for inefficiency is present as long as some compensation has to be made.

Third — and in some contrast to the preceding remarks — observe that the outcome cannot be "too inefficient". For if the degree of efficiency loss is high, then a proposer will be able to generate an efficiency improvement even with equal division, and such a proposal will be acceptable to all.

Fourth — and this follows up on the third point — notice that inefficient outcomes in symmetric games *must be asymmetric*. The reason is that a symmetric payoff outcome can be Pareto-dominated by equal division of the payoffs accruing to the grand coalition, as long as we're willing to grant that the game in question is grand-coalition-superadditive. It certainly is so in the Cournot example under consideration here: the joint monopoly payoff beats the sum of all equilibrium payoffs arising from any other coalition structure. Therefore symmetric inefficient outcomes are impossible: they can be trumped by equal-division proposals to the grand coalition.

Finally, where there is inefficiency — especially in a world of binding agreements — there is the scope for renegotiation. We have already discussed this issue when motivating the study of irreversible commitments, and the reader who wishes to refresh her memory can consult Chapter 2, Section 2.6. As mentioned there, we do drop the irreversibility assumption in later chapters.

6.2 Public Goods

We continue with another textbook example: the voluntary provision of public goods.[3] The twist is that we supplement the noncooperative voluntary contributions story with the ability to write binding agreements. The model that follows is one in which agents attempt to control a public *bad*, and can form coalitions that pledge cooperation across their members.

6.2.1 Model. Suppose that there are n regions (firms, countries, geographical centers) Each region can produce a public good — pollution control — the benefits of which accrue equally and additively to *all* regions. To generate benefit z involves a cost of $c(z)$ (increasing and strictly convex). In short, if Z is the total amount of pollution control produced by all regions, then the payoff to some region who produces z of it is

$$(6.2) \qquad Z - c(z).$$

Our first task is to describe the partition function for any coalition structure. We've made this very easy by assuming that all benefits are additive, so that the optimal actions of a coalition are unaffected by the ambient coalition structure. (Of course, this isn't true of the coalition's *payoffs*.) In particular, a coalition of s regions will surely produce pollution control of z per member, where z solves

$$(6.3) \qquad \max_z sz - c(z).$$

Let $z(s)$ be the solution. Let $f(s) \equiv sz(s)$, and $h(s) \equiv c(z(s))$. Thus, $f(s)$ denotes the optimal, aggregate output of coalition s, and $h(s)$ the corresponding per-member cost of provision of the public good. Then the net "internal" per-capita payoff is $g(s) \equiv f(s) - h(s)$. So if $\mathbf{n} = \{s_1, \ldots, s_m\}$ is a numerical structure, the average worth of s_i is just

$$(6.4) \qquad \alpha(s_i, \mathbf{n}) \equiv \sum_j f(s_j) - h(s) = g(s) + \sum_{j \neq i} f(s_j).$$

The second term above captures this average payoff as the sum of the "internal" and "external" payoffs.

Multiplying average worth by s_i yields precisely the partition function for the game.

---

[3]The analysis in this section is based on Ray and Vohra (2001).

As a special case, suppose that the cost function is quadratic: $c(z) = \frac{1}{2}z^2$. In this case, it is easy to see that a coalition of size s will set its total production $f(s)$ equal to s^2, and will incur a per-region cost $h(s)$ equal to $\frac{1}{2}s^2$ in doing so. It follows that

(6.5) $$\alpha(s_i, \mathbf{n}) = \sum_{j=1}^{m} s_j^2 - \frac{1}{2}s_i^2.$$

In all of what follows, we will assume that per-capita output $z(s)$ is increasing in s and that aggregate output $f(s)$ has nondecreasing increments.[4] The former condition is obviously self-explanatory and mild. It holds for all strictly convex and differentiable cost functions satisfying the end-point condition $c'(0) < 1$. The second holds widely as well; for instance, it is satisfied for all convex, constant-elasticity cost functions.[5]

6.2.2 Intuitive Approach. Study the quadratic case described in (6.5). To begin with, suppose that there are only two regions. Note that the stand-alone payoff to each region is 1.5, while if the two regions form a coalition, then the per-region payoff is 2 (simply apply (6.5)). It should therefore be obvious that any region will wish to team up with the other (and that such an offer will be accepted). Thus the two-region scenario implies full cooperation and efficiency.

The three-region case presents the first nontrivial prediction problem. If a single region contemplates staying on its own, it must predict what the remaining regions will do. But we know that in this case the remaining two regions *will* form a single coalition, so that the average worth of being alone is 4.5. A two-region coalition in this setting would average only 3. Finally, a three-region coalition averages 4.5 as well. This suggests that a region is indifferent between being on its own and being a member of a three-region coalition. Let's break this indifference in favor of the larger coalition (this assumption will not be needed in the formal analysis). So the three-region scenario is also conducive to efficiency.

Now turn to the four-region problem. By a similar process of computation and prediction, it turns out that if a region stays on its own, then it is in the interest of the other three regions to form

---

[4]That is, $f(s + 2) - f(s + 1) \geq f(s + 1) - f(s)$ for all s.
[5]Our results actually rely on the even weaker restriction that $g(s) + f(n - s)$ has nondecreasing increments. This is weaker because g always has this property.

a single coalition. Therefore the average worth of the stand-alone region is 9.5. In contrast, the formation of the grand coalition of four regions yields an average worth of only 8. This suggests that full cooperation cannot occur when there are four regions. Indeed, the numerical coalition structure that finally obtains is $(1, 3)$.

Now — strikingly enough — the five-region problem yields full cooperation. If one region were to stand alone, it would not be able to ensure the stability of the remaining four regions, which would configure themselves into the $(1, 3)$ structure. Consequently, the average worth of the original region must be 10.5, whereas the formation of the grand coalition yields 12.5.[6]

However — and now we start departing from an analysis based on intuition alone — full cooperation can (thereafter) no longer be reached unless there are at least eight regions, and then not again until there are at least thirteen regions.[7] But if we had to proceed by brute-force methods beyond the five-region case, the problem would very quickly become intractable. The results of Chapter 5 yield much quicker methods, and they apply to a wide class of equilibria and bargaining protocols. We supplement these results by drastically simplifying the algorithm of that chapter for the problem at hand.

6.2.3 Using the Algorithm.
We first put on record that the algorithm in Chapter 5 applies here:

PROPOSITION 6.2. *The only equilibrium coalition structure common to all rejector-friendly protocols (as bargaining frictions vanish) is the algorithmically derived structure* $\mathbf{n}^*$.

The proof of this proposition will involve a verification of SNAW for the public goods model. But we go further in establishing a simple characterization of $\mathbf{n}^*$ for the problem at hand. The simplification thus achieved is quite significant, and will reward the little patience needed to absorb the definition, which now follows.

---

[6]Other possibilities are similarly ruled out. For instance, the formation of a two-region coalition would yield an average worth of 11 for the two regions, which again is lower.

[7]The obvious comparison to the Fibonacci sequence— suggested by Andrew Postlewaite — ends here: the next size supporting full cooperation is 20!

Let $\mathcal{T} = \{m_1, m_2, \ldots\}$ be an infinite collection of increasing positive integers, with $m_1 = 1$. For any integer $n \geq 2$ define the $\mathcal{T}$-*decomposition* of n as a finite collection $\mathbf{s}(n) \equiv (t_1, \ldots, t_k)$ of (possibly repeated) elements of $\mathcal{T}$ satisfying the following properties:

(1) t_k is the largest integer in $\mathcal{T}$ which is strictly smaller than n.

(2) For any $i \in \{1, \ldots k-1\}$, t_i is the largest integer in $\mathcal{T}$ no greater than $(n - \sum_{j=i+1}^{k} t_j)$.

In other words, the $\mathcal{T}$-decomposition $\mathbf{s}(n)$ is obtained by subtracting the largest integer in $\mathcal{T}$ (that is strictly smaller than n) from n, then subtracting the largest possible integer in $\mathcal{T}$ (no greater than the remainder) from the remainder, and so on. Notice that since $1 \in \mathcal{T}$, $\mathbf{s}(n)$ is well-defined and unique for any positive $n > 1$.

Now consider a special collection $\mathcal{T}^* = \{m_1, m_2, \ldots\}$ of positive integers with the property that $m_1 = 1$, and for each $i \geq 1$, m_{i+1} is the *smallest* integer such that $m_{i+1} > m_i$ and

$$(6.6) \qquad g(m_{i+1}) \geq \alpha(t_1, \mathbf{s}(m_{i+1})) = \sum_{i=1}^{k} f(t_i) - h(t_1)$$

where $(t_1, \ldots, t_k)$ is the $\mathcal{T}^*$ decomposition of m_{i+1}.

This looks complicated but it really isn't. In fact — despite the apparent circularity in the construction — it is very easy to verify that $\mathcal{T}^*$ is uniquely defined, and is computable recursively.[8]

For any positive integer n, define the *decomposition* of n — call it $\bar{\delta}(n)$ — to be equal to its $\mathcal{T}^*$-decomposition if $n \notin \mathcal{T}^*$, or the singleton set $\{n\}$ if $n \in \mathcal{T}^*$.

We're now in a position to simplify the computation for $\mathbf{n}^*$.

PROPOSITION 6.3. *The algorithmic coalition structure is just the decomposition of n:* $\mathbf{n}^* = \bar{\delta}(n)$.

In the standard equilibrium isolated by Proposition 6.2, coalitions that form attempt to maximize their average worth. This is how we obtain the algorithmic outcome $\mathbf{n}^*$. To understand the claim of

---

[8]The collection $\{m_1, \ldots, m_i\}$ is all that is needed to compute the $\mathcal{T}^*$-decomposition of any integer n that exceeds m_i, so recursion is possible. All we need to do is make sure that the inequality (6.6) *is* satisfied for some $m_{i+1} > m_i$. To see this simply note that $f(n) - h(n) \to \infty$ as $n \to \infty$.

Proposition 6.3 that this recursive procedure boils down to finding the decomposition of n, return to the quadratic example. It is easy to write down the first few terms in $\mathcal{T}^*$: these are $\{1, 2, 3, 5, \ldots\}$.

Now follow the procedure. To construct the equilibrium structure when there are, say, six regions, it will suffice to compare $\alpha(1, (1, 5))$, which is 25.5, to $\alpha(6, (6)))$, which is 18.[9] This shows that $(1, 5)$ is the equilibrium coalition structure when $n = 6$. It also means that $6 \notin \mathcal{T}^*$. Thus, when $n = 7$, we need only compare $\alpha(2, (2, 5))$, which is 27, to $\alpha(7, (7))$, which is 24.5. This establishes that $(2, 5)$ is the equilibrium coalition structure when $n = 7$, and, in particular, that $7 \notin \mathcal{T}^*$. Therefore, when $n = 8$, the comparison is to be made between $\alpha(3, (3, 5))$, which is 29.5, to $\alpha(8, (8))$, which is 32. This signals the return of the grand coalition: $8 \in \mathcal{T}^*$.

With this sort of argument, it is easy enough to proceed further, and to show that the next two elements in $\mathcal{T}^*$ are 13 and 20. Our decomposition achieves a significant reduction in the number of checks for an equilibrium coalition structure.

6.2.4 On Efficiency. Notice that efficient outcomes are to be had *only* through the formation of the grand coalition. We've already seen in the quadratic case that this happens sometimes, but not all the time. Yet, as in the Cournot example, the system cannot tolerate too great a loss in efficiency, for then the grand coalition would be proposed (and accepted) by all. Let's examine these ideas in some more detail.

The quadratic example hints at a feature that is different from the Cournot example. Recall in the latter example that once n exceeds 4, efficiency is *never* to be had. Yet the quadratic example for public goods suggests that efficiency is "periodically" restored along a subsequence of region sets. The generality of this observation (for the public goods problem) is immediately confirmed by Proposition 6.3. The set $\mathcal{T}^*$ is an *infinite* set, so there must be infinitely many values of n — precisely all the entries in $\mathcal{T}^*$ in fact — for which the decomposition of n is just $\{n\}$. So along an infinite subsequence of region populations, the outcome is fully efficient.

---

[9]According to our result, it is not necessary to compute $\alpha(2, (2, 4))$, $\alpha(2, (1, 2, 3))$ or the average worth of a region in any of the several other coalition structures, which saves on computation.

What about the (more numerous) cases in which the outcome fails to be efficient? In these situations the reasoning is broadly the same as in the Cournot model: the potential for counteroffers drives a wedge between private and social incentives. It also implies that the equilibrium coalition structure must be asymmetric: symmetric and inefficient outcomes must be dominated by the grand coalition using equal division of payoffs. This asymmetry is both of intrinsic and instrumental interest so let's take a closer look:

PROPOSITION 6.4. *Suppose that the grand coalition does not form in equilibrium, i.e.,* $\mathbf{n}^* = \bar{\delta}(n) = (n_1, \ldots, n_k)$, *where* $k \geq 2$. *Then* $n_i \neq n_j$ *for all* $i, j \in \{1, \ldots k\}$.

Unlike the Cournot model, the asymmetry here comes in a very strong flavor. *Every* equilibrium coalition must have a different size! The reason is that externalities are additive, so that every substructure of the equilibrium coalition structure is indeed an equilibrium of the game with player set equal to all the regions in that substructure. (The Cournot model does not have this property.) It follows that every substructure is asymmetric, and this can only happen if no two coalitions have the same size.

Now for the instrumental use of asymmetry. A corollary of Propositions 6.3 and 6.4 is that the largest coalition in the equilibrium coalition structure must have more than half the players of the game in it. For if it didn't in some n-player game, the largest number would appear at least *twice* in the decomposition of n, which contradicts Proposition 6.4. This observation places an immediate upper bound on just how inefficient things can get: the *equilibrium* social surplus must be at least as large as the *maximum* social surplus generated in a community of half the size. It also places a bound on how many different coalitions can form in equilibrium. Both these observations are summarized in

PROPOSITION 6.5. *The following assertions about efficiency hold:*

(i) *For each* n, *the ratio of equilibrium to potential surplus exceeds* $\frac{g(n/2)}{g(n)}$.

(ii) *If* k *is the number of equilibrium coalitions, then* $k < \log_2 n + 1$.

Ray and Vohra (2001, Theorem 2) actually establish a stronger (and more subtle) version of part (i): they show that the ratio must be at least $4g(n/2)/3g(n)$. We omit this argument here.

In the case of a quadratic cost function the proposition yields an "efficiency ratio" of at least 25% (or 33% if the Ray–Vohra result is used). Note that the bound is independent of n, the number of regions.

Finally, part (ii) is an immediate consequence of the discussion preceding the statement of the proposition. If there are k coalitions in a decomposition, and if the size of each successive coalition (counting from the largest) must exceed half the number of the remaining regions, then the total number of regions must exceed 2 raised to the power of the number of coalitions, which yields the required result. Note that this bound predicts substantial cooperation: for instance, no more than seven coalitions can form when there are a hundred negotiating regions (or countries).

6.3 Proofs

Proof of Proposition 6.1. Our first step is to calculate $t(\mathbf{n})$ as generated by the algorithm.

For each substructure $\mathbf{n}$, let $R(\mathbf{n})$ denote the number of coalitions in $\mathbf{n}$ and let $m(\mathbf{n})$ be the number of "active players" who remain "outside" $\mathbf{n}$.[10] Sometimes we will simply call these numbers R and m when the context is perfectly clear.

STEP 1. $t(\mathbf{n}) = 1$ if $m(\mathbf{n}) < (R(\mathbf{n}) + 1)^2$.

Proof. By induction on m. The result is trivially true for $m = 1$ and any R such that the above inequality holds (which is all $R \geq 1$). Now fix any $m > 1$. Suppose recursively that the claim is true for $m - 1$ and less.

Consider any $\mathbf{n}$ such that $m(\mathbf{n}) = m$ and $R = R(\mathbf{n})$ satisfies $m < (R + 1)^2$. For any choice of t, the induction assumption clearly guarantees that the structure of singletons will form thereafter. Consequently the average worth of t is given by

$$\frac{v(t, c(\mathbf{n}.t))}{t} = \frac{1}{t[R + 1 + (m - t) + 1]^2} = \frac{1}{t[R + (m - t) + 2]^2}.$$

Differentiate the expression $t[R + (m - t) + 2]^2$ with respect to t. It is easy to see that for low values of t the derivative is positive while for all higher values the derivative is negative. So the *reciprocal* of

---

[10]In other words, $m(\mathbf{n}) = n - K(\mathbf{n})$, where $K(\mathbf{n})$ was defined in Section 5.2.

this expression (which is average worth) is U-shaped in t. Thus to find the maximum, it suffices to simply compare $t = 1$ with $t = m$. Do so; we see that $t(\mathbf{n}) = 1$ if and only if

(6.7) $(R + m + 1)^2 < m(R + 2)^2.$

Manipulate this to see that the equivalent condition is

(6.8) $(R + 1)^2 > m$

which completes the inductive proof of Step 1.

STEP 2. $t(\mathbf{n}) = m$ if $(R(\mathbf{n}) + 1)^2 \le m(\mathbf{n}) < (R(\mathbf{n}) + 2)^2 + 1.$

Proof. Consider the choice of t to maximize average worth at this stage. Note that for every $t \ge 1$, if we define $R' \equiv R+1$ and $m' \equiv m-t$, then the given condition of this step

$$m < (R + 2)^2 + 1$$

assures us that

$$m' < (R' + 1)^2,$$

so that the subsequent structure after choosing t will all be singletons (using Step 1). So the comparison is just as in Step 1; it is a choice between $t = 1$ and $t = m$. But this time, given that

$$m \ge (R + 1)^2,$$

$t(\mathbf{n}) = m$ must be the algorithmic choice.

STEP 3. $t(\mathbf{n}) = 1$ if $m(\mathbf{n}) \ge (R(\mathbf{n}) + 2)^2 + 1.$

Proof. Omitted. Mechanical; various subcases need to be considered (see Bloch (1996) for details).

Given this description of $t(.)$, we first verify that $\mathbf{n}^*$ is of the form described in the statement of the theorem. Starting at ϕ, Step 3 dictates that singletons must form (so that $t(\mathbf{n}) = 1$ and $R(\mathbf{n})$ equals the number of elements of $\mathbf{n}$ for all such $\mathbf{n}$) until we reach the *first* nonnegative integer L such that

(6.9) $n - L < (L + 2)^2 + 1.$

At this stage it is easy to see that the condition of Step 2 holds exactly; that is, L also satisfies

$$n - L \ge (L + 1)^2.$$

For if not, then $n - L < (L + 1)^2$. This means that L is a positive integer, so that $L' \equiv L - 1$ is a nonnegative integer. But then, $n - L' < (L' + 1)^2 + 1$, which contradicts the definition of L.

To complete the proof, we must verify that average algorithmic worth is strongly nonincreasing.

To do so, we note that starting from any $\mathbf{n}$, the algorithm guarantees that larger coalitions always form later (if they do at all) than smaller coalitions. It immediately follows that average worth is weakly nondecreasing.

Now suppose $t(\mathbf{n}) = 1$. In such cases there is nothing further to be proved since restricted maximizers cannot be distinct from the unrestricted maximizer $t = 1$.

It remains to consider the case in which $t(\mathbf{n}) = m(\mathbf{n}) > 1$. In this case, we claim that the only other restricted maximizer is $t = 1$. Obviously, it will suffice to show this for the case in which the upper bound for the maximization problem is $m(\mathbf{n}) - 1$. (For in that case $t = 1$ must be the restricted maximizer for all other bounds as well.)

To prove this, note — just as in Step 2 — that no matter what coalition forms at this stage, the remaining coalition structure will be one of singletons. So the comparison is just as in Step 1; compare "end-point" choice $t = 1$ and — remembering that we are restricted by the bound $m - 1$ — the choice $t = m - 1$. That is, we substitute $t = 1$ and then $t = m - 1$ in the expression $t[R + (m - t) + 2]^2$ and choose the one with the lower value (remember this expression is the reciprocal of payoffs). So $t = 1$ if

$$[R + (m - 1) + 2]^2 = [(R + 1) + (m - 1) + 1]^2 < (m - 1)[(R + 1) + 2]^2.$$

This expression is exactly analogous to (6.7) in Step 1 with $m - 1$ in place of m and $R + 1$ in place of R. So, invoking (6.8), the condition equivalent to this is

$$[(R + 1) + 1]^2 > (m - 1),$$

or

$$m < (R + 2)^2 + 1,$$

which is certainly the case because the *unrestricted* maximizer is m (consult the conditions in Step 2).

So $t = 1$ is the restricted maximizer, and thereafter we have the coalition structure of singletons. But the worth after $t = 1$, $a(\mathbf{n}.1)$, is then just the payoff to a singleton, which is no higher than the average worth of the grand coalition at **n**. By Step 2 this attains the highest average worth $a(\mathbf{n})$, and so $a(\mathbf{n}) \geq a(\mathbf{n}.1)$. Now we have verified the required property, so that all equilibria of the game are described by the algorithm. □

We turn now to proofs for the public goods model. It will be useful to prove Propositions 6.2 and 6.3 in reverse order.

First, we collect some elementary observations regarding the functions f and g.

LEMMA 6.1. (i) If $g(t) + f(n-t) \geq g(s) + f(n-s)$ for some $1 \leq s \leq t < n$, then $g(t') + f(n-t') \geq g(t) + f(n-t)$ for all $t \leq t' < n$.

(ii) If $t \geq s$, then $g(s) + f(t) \geq g(t) + f(s)$, with strict inequality whenever $t > s$.

Proof. By our assumption on f, $g(s) + f(n-s)$ has nondecreasing increments on $[0, n]$. If such a function is nondecreasing over some range of s, it can never decrease thereafter; this proves (i).

To prove (ii), observe that $h(s)$ is increasing in s (because $z(s)$ is, by assumption). It follows that if $t \geq s$, then

$$g(s) + f(t) = f(s) + f(t) - h(s) \geq f(s) + f(t) - h(t) = g(t) + f(s),$$

with strict inequality whenever $t > s$, which proves (ii). □

The next lemma collects some elementary observations regarding decompositions. For any positive integer n, define its *strict decomposition* $\mathbf{s}(n)$ to be just the $\mathcal{T}^*$-decomposition of n. (So $\bar{\delta}(n) = \mathbf{s}(n)$ if $n \notin \mathcal{T}^*$.)

Use notation such as $s.\bar{\delta}$ or $t.s.\mathbf{s}$ to denote the numerical structures generated by putting together s and/or t with a decomposition. (The order of the integers is unimportant.) As a notational convention, $\bar{\delta}(0) = \emptyset$ and $s.\emptyset)$ is just s.

LEMMA 6.2. *For any positive integer* n,

(i) If $\bar{\delta}(n) = t_1.\ldots.t_k$, then for all $t \in \{1, \ldots, t_1\}$, $\bar{\delta}(n-t) = \bar{\delta}(t_1-t).t_2.\ldots.t_k$.

(ii) If t_k is the largest value in $\mathbf{s}(n)$ then $\bar{\delta}(m) = \bar{\delta}(m - t_k).t_k$ for all m such that $t_k \leq m < n$.

(iii) $\mathbf{s}(n) = \bar{\delta}(n - t).t$, whenever $t \in \mathbf{s}(n)$.

Proof. These observations follow directly from the definition of a decomposition. □

The following lemma is central to the main argument.

LEMMA 6.3. *Suppose that n has strict decomposition $\mathbf{s}(n) = \{t_1, \ldots, t_k\}$. Then*

$$(6.10) \qquad \alpha(t_1, \mathbf{s}(n)) \geq \alpha(t, t.\bar{\delta}(n - t)) \text{ for all } t \in \{1, \ldots, t_1 - 1\},$$

$$(6.11) \qquad \alpha(t_1, \mathbf{s}(n)) > \alpha(t, t.\bar{\delta}(n - t)) \text{ for all } t \in \{t_1 + 1, \ldots, n - 1\}.$$

Proof. Proceed by induction. Clearly for $n = 2$ the assertion is trivially true because both the sets in question are empty. Suppose, then, that the lemma is fully established for all integers m that lie between 2 and $n - 1$, and consider the lemma for n.

First take $t \in \{1, \ldots, t_1 - 1\}$. Then $\bar{\delta}(n - t) = \bar{\delta}(t_1 - t).t_2.\ldots.t_k$ (Lemma 6.2, part (i)), so that

$$\alpha(t, t.\bar{\delta}(n - t)) = \alpha(t, t.\bar{\delta}(t_1 - t).t_2.\ldots.t_k)$$

$$= \alpha(t, t.\bar{\delta}(t_1 - t)) + \sum_{j=2}^{k} f(t_j)$$

$$\leq \alpha(s_1, \mathbf{s}(t_1)) + \sum_{j=2}^{k} f(t_j)$$

$$\leq f(t_1) - h(t_1) + \sum_{j=2}^{k} f(t_j)$$

$$= \alpha(t_1, \mathbf{s}(n)),$$

where s_1 is the first term in the strict decomposition[11] of t_1, and the inequality in that line holds by the induction hypothesis. The second inequality holds using (6.6) and the fact that $t_1 \in \mathcal{T}^*$.

Next, take $t \in \{t_1 + 1, \ldots, n - 1\}$. There are now two subcases, each of which we consider in turn.

SUBCASE 1. $t_1 < t \leq n - t_k$. Note that in this case $k \geq 3$. Using Lemma 6.2, part (ii), the decomposition of $n - t$ is given by $\bar{\delta}(n - t_k - t).t_k$, so

---

[11]This is well defined because in the case under consideration, t_1 must be at least 2.

that

$$\alpha(t, t.\bar{\delta}(n-t)) = \alpha(t, t.\bar{\delta}(n-t_k-t).t_k)$$
(6.12)
$$= \alpha(t, t.\bar{\delta}(n-t_k-t)) + f(t_k).$$

Since $k \geq 3$, it follows that $\bar{\delta}(n-t_k) = s(n-t_k)$ and, by Lemma 6.2, part (iii), $\bar{\delta}(n-t_k) = t_1 \ldots t_{k-1}$. Applying the induction hypothesis to the integer $n - t_k$, we see that

(6.13) $$\alpha(t, t.\bar{\delta}(n-t_k-t)) < \alpha(t_1, s(n-t_k)).$$

Combining (6.12) and (6.13) and using Lemma 6.2 part (iii) once again, we may conclude that

$$\alpha(t, t.\bar{\delta}(n-t)) < \alpha(t_1, s(n-t_k)) + f(t_k) = \alpha(t_1, s(n))$$

which completes the proof in this subcase.

SUBCASE 2. $t > n-t_k$. Suppose (6.11) does not hold for some $t > n-t_k$, i.e.,

(6.14) $$\alpha(t, t.\bar{\delta}(n-t)) \geq \alpha(t_1, s(n)).$$

Recall that $g(t) \equiv f(t) - h(t)$ for all t. Notice that the highest possible average worth to a coalition of size t arises when the rest of the regains form one single coalition, i.e., $\alpha(t, t.\bar{\delta}(n-t)) \leq g(t) + f(n-t)$. Combining this information with (6.14), we see that

(6.15) $$g(t) + f(n-t) \geq \alpha(t_1, s(n)).$$

Recalling the previous subcase (or the definition of $s(n)$ in case $n = t_1 + t_k$), and noting that $\bar{\delta}(t_k) = t_k$, it follows that

(6.16) $$\alpha(t_1, s(n)) \geq \alpha(n-t_k, (n-t_k).t_k) = g(n-t_k) + f(t_k).$$

Combining (6.15) and (6.16), we have

$$g(t) + f(n-t) \geq g(n-t_k) + f(t_k).$$

Applying Lemma 6.1, part (i), we conclude that

(6.17) $$g(n-1) + f(1) \geq g(t) + f(n-t) \geq g(n-t_k) + f(t_k).$$

Applying Lemma 6.1, part (ii), to the first and third expressions of the inequality (6.17), we conclude that $n - t_k > 1$. This means that

$n - 1 \notin \mathcal{T}^*$. In other words, we can write $\bar{\delta}(n-1) = (s_1, \ldots, s_q)$ where $q \geq 2.$[12] Using (6.6), we conclude that

$$g(n-1) < \alpha(s_1, \bar{\delta}(n-1)) = g(s_1) + \sum_{j=2}^{q} f(s_j).$$

Since we know from Lemma 6.1, part (ii), that $g(1) + f(s_1) \geq g(s_1) + f(1)$, this inequality can be rewritten as

$$g(n-1) + f(1) < g(1) + \sum_{j=1}^{q} f(s_j)$$
$$= \alpha(1, 1.\bar{\delta}(n-1))$$
$$\leq \alpha(t_1, \mathbf{s}(n)).$$

But this, along with (6.17), implies

$$\alpha(t_1, \mathbf{s}(n)) > g(t) + f(n-t),$$

which contradicts (6.15). □

Proof of Proposition 6.3. When the number of regions is n, use the notation $T(n)$ to denote the algorithmic choice of coalition size at the initial stage (this corresponds to $t(\emptyset)$ in Chapter 5). Use the notation $\mathbf{C}(n)$ to refer to algorithmic coalition structure $\mathbf{n}^*$. (Employ the convention that $\mathbf{C}(0)$ is empty.) We are to show that $\mathbf{C}(n) = \bar{\delta}(n)$.

The proposition is trivially true when $n = 1$, so assume inductively that for some integer $n \geq 2$, $\mathbf{C}(m) = \bar{\delta}(m)$ for all $m = \{1, \ldots, n-1\}$. All we need to show is that the algorithmic choice $T(n)$ is the first term in the decomposition of n. In other words, if $\bar{\delta}(n) = t_1 \ldots t_k$, we need to prove that

(6.18) $\qquad t_1 = \max\{\arg\max_{t \in \{1,\ldots,n\}} \alpha(t, t.\mathbf{C}(n-t))\}.$

Let $\mathbf{s}(n) = s_1 \ldots s_q$. From Lemma 6.3 we know that

(6.19) $\qquad s_1 = \max\{\arg\max_{t \in \{1,\ldots,n-1\}} \alpha(t, t.\mathbf{C}(n-t))\}.$

Consider the two following cases:

1. Suppose $s_1 = t_1$. This means that n does not satisfy (6.6), i.e.,

$$\alpha(t_1, t_1.\bar{\delta}(n-t_1)) > \alpha(n, n).$$

---

[12] In fact we know that if $t_1 = 1$, then $\bar{\delta}(n-1) = (t_2, \ldots, t_k)$ and if $t_1 > 1$, then $\bar{\delta}(n-1) = (\bar{\delta}(t_1 - 1).t_2, \ldots, t_k)$. But this need not concern us in what follows.

Since, by the induction hypothesis, $C(n - t_1) = \bar{\delta}(n - t_1)$, and $s_1 = t_1$, this along with (6.19) implies (6.18).

2. Suppose $t_1 > s_1$. This means that $t_1 = n$ and that n satisfies (6.6) holds, i.e.,

$$\alpha(t_1, \bar{\delta}(n)) = \alpha(n, n) \geq \alpha(s_1, s_1.\bar{\delta}(n - s_1)),$$

and again (6.19) implies (6.18). □

We now turn to the proof of Proposition 6.2. By virtue of Proposition 5.4 in Chapter 5, it will suffice to establish SNAW. To this end, for any positive integer n, say that $t \in \{1, \dots, n - 1\}$ is a *constrained maximizer* if

(6.20) $\qquad \alpha(t, C(n - t)) \geq \alpha(t', C(n - t'))$ for all $t' \in \{1, \dots, t\}$.

LEMMA 6.4. *For any positive integer n, let $\phi(n)$ be the smallest integer in $s(n)$, and let ϕ^k denote the k-fold composition of ϕ. Then, if t is a constrained maximizer, $t = \phi^k(n)$ for some integer k.*

Proof. Suppose $t \neq \phi^k(n)$ for any integer k. Noting that $\phi^k(n) = 1$ after some finite k, let s be the largest integer of the form $\phi^k(n)$ such that s is smaller than t. Use the convention $\phi^0(n) = n$. Then

$$s = \phi^k(n) < t < \phi^{k-1}(n) \equiv m.$$

Using Lemma 6.2 and the fact that $m = \phi^{k-1}(n)$,

(6.21) $\qquad \bar{\delta}(n - t') = \bar{\delta}(m - t').\bar{\delta}(n - m)$ for all $t' < m$.

Since s is the first term in the strict decomposition of m, it follows from Lemma 6.2 that

$$s.\bar{\delta}(n - s) = s.\bar{\delta}(m - s).\bar{\delta}(n - m) = s(m).\bar{\delta}(n - m).$$

Thus,

$$
\begin{aligned}
\alpha(s, s.\bar{\delta}(n - s)) &= \alpha(s, s(m - s)) + \sum_{t' \in \bar{\delta}(n-m)} f(t') \\
&> \alpha(t, t.\bar{\delta}(m - t)) + \sum_{t' \in \bar{\delta}(n-m)} f(t') \\
&= \alpha(t, t.\bar{\delta}(n - t)),
\end{aligned}
$$

where the inequality uses (6.11) of Lemma 6.3 applied to the integer m. Since $C(n) = \bar{\delta}(n)$ for every n, we may conclude that

$$\alpha(s, s.C(n - s)) > \alpha(t, t.C(n - t)),$$

so that t fails (6.20) and cannot, therefore, be a constrained maximizer. □

Proof of Proposition 6.2. For any positive integer n, define

$$\alpha^*(n) \equiv \max_t \alpha(t, t.\mathbf{C}(n - t)) = \alpha(T(n), T(n).\mathbf{C}(n - T(n))).$$

To establish SNAW, it will suffice to prove that if t is a constrained maximizer, then

(6.22) $$\alpha^*(n) \geq \alpha^*(n - t) + f(t).$$

The reason is simple. Interpret n as the "remaining set" of players (some substructure has left). Because all externalities are additive and and the good is purely public, $\alpha^*(n)$ is just average algorithmic worth minus these externalities. For $n-t$, $\alpha^*(n-t)+f(t)$ is *also* average algorithmic worth minus the *very same externalities*, but including the external effect of the just-formed coalition t. This is why (6.22) is tantamount to establishing the SNAW condition (5.5) in Chapter 5.

So to establish (6.22), then, note that by Lemma 6.4, $t = \phi^k(n)$ for some $k \geq 1$. Let $m = \phi^{k-1}(n)$ (to be interpreted, by convention, as n if $k = 1$). Then, if s_2 is the *second* term in the strict decomposition of m (t being the first),

$$\alpha(t, \mathbf{s}(m)) > \alpha(s_2, \mathbf{s}(m)),$$

by virtue of the simple fact that $t < s_2$ and by Lemma 6.1, part (ii). Using (6.21), we have:

$$\alpha(t, t.\bar\delta(n - t)) = \alpha(t, \mathbf{s}(m).\bar\delta(n - m)) > \alpha(s_2, \mathbf{s}(m).\bar\delta(n - m)).$$

Since $\mathbf{s}(m) = t.\bar\delta(m - t)$, we can appeal to (6.21) to assert that

$$\alpha(s_2, \mathbf{s}(m).\bar\delta(n - m)) = \alpha(s_2, t.\bar\delta(m - t).\bar\delta(n - m)) = \alpha(s_2, \bar\delta(n - t)) + f(t).$$

Since $\mathbf{C}(n) = \bar\delta(n)$ for all n, it follows from the definition of α^* that

$$\alpha(s_2, \bar\delta(n - t)) = \alpha^*(n - t).$$

Combining the last three equations, we see that

(6.23) $$\alpha(t, t.\bar\delta(n - t)) > \alpha^*(n - t) + f(t).$$

On the other hand, the definition of α^* implies that

(6.24) $$\alpha^*(n) \geq \alpha(t, t.\mathbf{C}(n - t, T)) = \alpha(t, t.\bar\delta(n - t)).$$

Combining (6.23) and (6.24), we obtain (6.22). □

Proof of Proposition 6.4. Suppose there exist $n_i, n_j \in \bar{\delta}(n)$ such that $i \neq j$ and $n_i = n_j = m$. Then it is easy to see that $\bar{\delta}(2m) = (m, m)$. In particular, $2m \notin \mathcal{T}^*$, so that (6.6) is negated, which means that

$$(6.25) \qquad\qquad f(m) + g(m) > g(2m).$$

Recall that $f(m) = mz(m)$, where $z(m)$ is the solution to (6.3) for $s = m$, and $f(m) + g(m) = 2mz(m) - c(z(m))$. But $g(2m)$, by definition, equals the *maximum value* of $2mz - c(z)$ over all z, so that (6.25) fails, a contradiction. □

As discussed in the text, the proof of Proposition 6.5 follows quite easily from Proposition 6.4, and so is omitted. For a proof of the extended result mentioned in the main text, see the argument that establishes Theorem 2 in Ray and Vohra (2001).

6.4 Summary

This chapter studies two applications of the theory developed for symmetric partition functions.

First, we study a Cournot oligopoly. By verifying the appropriate condition from Chapter 5, we are allowed to apply to algorithm of that chapter to this case. We verify that once the number of firms passes a critical threshold, the outcome is invariably inefficient for the firms. Even though the formation of the grand coalition is a potential binding agreement which is efficient, the equilibrium cartel structure in the industry typically consists of one large firm, along with several stand-alone firms.

In the second application, we study binding agreements in a pollution control problem. Once again, equilibrium agreements and coalition structures may be characterized by first verifying the conditions in Chapter 5 and then applying the algorithm developed there. The outcome here is somewhat different from the Cournot counterpart: while it is typically inefficient, there is a subsequence of agent populations along which an efficient outcome results.

These differences notwithstanding, both the applications have the following features in common: (a) they serve as evidence for the generality of the conditions (NAW and SNAW) used in Chapter 5; (b) they generate inefficient outcomes despite the ability to write any binding agreement, in principle; (c) they show why inefficient outcomes in symmetric games typically involve asymmetric coalition structures; and (d) they reveal that the degree of inefficiency cannot be "too high", in a sense made clear in the discussion above.

CHAPTER 7

Irreversible Agreements: The General Case

The purpose of this chapter is to make progress on the important question of coalition formation when agents are heterogeneous. It is hardly necessary to motivate such an exercise. Symmetric games are perfectly fine for a host of different applications. But there are several situations in which the relevant agents differ in ability, wealth or power, and we would like to be able to say something about them.

The analysis that follows is necessarily complex. Not only are we attempting to deal with heterogeneous agents, but we are simultaneously in search of a general model that will incorporate cross-coalitional externalities. This is why we proceed in two main steps. In the first step we banish the externalities but allow for arbitrary variation across players. In the second step we bring the externalities back.

The first step is actually useful beyond its role as a transition device. It is also the stuff of classical cooperative game theory, built on the characteristic function. It therefore serves as a way of benchmarking our results to the traditional literature. Moreover — my reservations about characteristic functions notwithstanding — there is little doubt that they fit well in a variety of situations.[1] Accordingly, we begin with a quick survey of traditional cooperative game theory, albeit in highly limited and selective form.

---

[1] My reservations had to do with twisting other entirely incompatible situations to fit the characteristic function viewpoint.

Throughout this chapter, we also restrict ourselves to transferable payoffs. Unlike symmetric games, a full extension to the NTU case when players are heterogeneous is not immediate. See Section 14.8 in Chapter 14 for more discussion.

7.1 Characteristic Functions and the Core

Chapter 2 (Section 2.2) already introduced the *characteristic function*, a key concept in the theory of cooperative games. This will also be our starting point for this chapter. Recall that a (transferable utility) characteristic function is just a partition function in which partitions play no role: each coalition S has a well-defined *worth* $v(S)$ which can be divided freely among its members. We have already mentioned several examples which can be usefully captured as a characteristic function game.

A central equilibrium notion in cooperative game theory is that of the core. Consider the grand coalition N. Say that a feasible allocation **u** for N is *blocked by coalition S* if $v(S) > \sum_{i \in S} u_i$. The *core* of a characteristic function is the set of all unblocked allocations.

To be sure, there is no particular reason to focus on the grand coalition in this definition. After all, the game itself may not be "physically superadditive", and only subcoalitions may have any shot at viability. We've discussed this possibility in Chapter 3 and will do so again in Chapter 10. In a game with irreversible commitments, however, there is no loss of generality in the assumption of "effective superadditivity": a larger coalition has among its cooperative options the *option* to break up into smaller subgroups. This notion of the superadditive cover will be problematic when negotiations are ongoing (see Chapter 10) but for now we adopt it. The definition is standard: if S and T are disjoint coalitions, then

$$v(S \cup T) \geq v(S) + v(T).$$

With the superadditivity assumption in place, we may — without loss of generality — look only to the grand coalition for efficient payoff allocations. However, it is well known that superadditivity isn't good enough for a nonempty core:

EXAMPLE 7.1. *Let $N = \{123\}$, $v(i) = 0$, $v(\{ij\}) = a$ for all i and j, $v(N) = b$ and assume that $a > 0$ and $b > a$. Then the game is superadditive. But its core is nonempty only under the stronger restriction that $b \geq 3a/2$.*

For suppose that the core is nonempty. Let **y** be a core allocation. Then

$$y_i + y_j \geq a$$

for all i and j. Adding this up over the three possible pairs, we see that

$$2(y_1 + y_2 + y_3) \geq 3a,$$

or $b \geq 3a/2$.

This argument is easy enough to generalize. As in the case of partition functions, note that a characteristic function is symmetric if the worth of each coalition is expressible as a function of the number of players in that coalition. As before, we abuse notation slightly to write $v(S)$ as $v(s)$, where s is the cardinality of S.

OBSERVATION 7.1. *The core of a symmetric TU game is nonempty if and only if*

(7.1)
$$\frac{v(n)}{n} \geq \frac{v(s)}{s}$$

for every coalition of size s.

Proof. To see that (7.1) is sufficient simply use equal division and show that it is a core allocation. For the necessity of (7.1), use the obvious extension of the argument in the example above. □

Of course, this elementary proposition can (and has) been further generalized. For transferable utility characteristic functions the well-known Bondareva–Shapley theorem provides a necessary and sufficient condition for existence, using a generalization of the "average worth" condition (7.1) known as balancedness. But it is clear already that superadditivity, and the ability of a group to seize upon an efficient allocation, are two properties that need not coincide.

At the same time, does the notion of a core — empty or otherwise — help us in understanding the agreements that might be written "in equilibrium", and the attendant coalition structures that might arise? The answer is: it helps, but not by much.

First, if the core is empty, *something* still has to happen, presumably. Perhaps a structure of subcoalitions forms, or perhaps an allocation for the grand coalition still comes about, on the grounds that any attempt to block the allocation will be "further" blocked in some way. In Part 3 of this book, which studies the "blocking approach" to coalition formation, I try to take these matters further.

Second, one must observe that once these additional routes open up, there is no guarantee that a core allocation will be implemented *even if* the core is nonempty. The broader possibilities described in the previous paragraph may still be pertinent even when the core is nonempty.

Third, the concept of the core *presumes* that a solution (if nonempty) must be efficient. The previous chapters display ominous signs that this may not be the case.

Finally, the core cannot convincingly take us beyond characteristic functions, even though attempts have been made in this direction (see, e.g., Lucas (1963)). With partition functions, a coalitional block cannot be suitably defined unless the ambient coalition structure is fully described.

The purpose of this brief review was twofold. First, it acquainted us with characteristic functions and a classical solution concept, the core. Second, it will serve to place the analysis that follows in this traditional perspective.

7.2 Equilibrium Response Vectors

We apply the model of coalitional bargaining introduced in Chapter 4. Throughout this chapter, we restrict ourselves to the rejector-proposes protocol. As we discuss in Section 7.7 below, the broad methodology will apply over a wider range of protocols, though the general territory is admittedly unexplored at the time of writing this monograph. Because the analysis here is harder going as it is, expositional tractability suggests that I attempt to uncover some of the features of the heterogeneous model without having to simultaneously deal with an entire class of protocols.

I begin by developing the idea of an *equilibrium response vector*, a player-specific collection of thresholds above which players will (loosely speaking) accept the offers made to them.

Fix a stationary Markovian equilibrium. Let $x_i(S, \delta)$ be the equilibrium payoff to player i when i is the proposer and S is the set of active players. For each i, define

$$y_i(S, \delta) = \delta x_i(S, \delta).$$

Then, if j were to receive a proposal $(T, \mathbf{z})$ at player set S such that $z_k > y_k(S, \delta)$ for all $k \in T$ yet to respond, including j, then j must accept.[2] On the other hand, if this inequality held for all k (yet to respond), barring j for whom the opposite strict inequality holds, then j must reject.[3] I therefore use the terminology *equilibrium response vector* to describe the collection $\mathbf{y}(S, \delta)$.

The next observation is basic to all that follows:

OBSERVATION 7.2. *For every* (S, δ) *and each* $i \in S$,

$$(7.2) \qquad y_i(S, \delta) \geq \delta \max_{T: i \in T \subseteq S} \left[v(T) - \sum_{j \in T-i} y_j(S, \delta) \right].$$

with equality holding whenever i makes a proposal which is accepted.

The proof is pretty trivial. By making an offer $\mathbf{z}$ to any coalition T (with $i \in T$) such that $z_j > y_j(S, \delta)$ for every $j \in T$, $j \neq i$, i can guarantee acceptance of the proposal. It follows that

$$x_i(S, \delta) \geq \max_{T: i \in T \subseteq S} \left[v(T) - \sum_{j \in T-i} y_j(S, \delta) \right],$$

and using $y_i(S, \delta) = \delta x_i(S, \delta)$, we are done with (7.2).

Notice, furthermore, that if (7.2) holds with strict inequality for some i, then i *must* be making an equilibrium proposal $(T, \mathbf{z})$ such that $z_j < y_j(S, \delta)$ for some $j \in T$. Look at the last player in T's response order for which this inequality holds. That player must reject — assuming the proposal makes it that far — so the proposal cannot be acceptable. □

We've been careful about distinguishing between acceptable and unacceptable proposals in equilibrium. Might unacceptable proposals ever be made in equilibrium? They might.

EXAMPLE 7.2. $N = \{1234\}$, $v(\{1j\}) = 50$ *for* $j = 2, 3, 4$, $v(\{ij\}) = 100$ *for* $i, j = 2, 3, 4$, *and* $v(S) = 0$ *for all other* S.

---

[2]This follows from a simple inductive argument which uses the assumption that responses are sequential.

[3]The other possibilities have unclear implications at this stage. For instance, if $z_j < y_j(S, \delta)$ but it is *also* the case that $z_k < y_k(S, \delta)$ for a later respondent k, should j reject? Unclear. Indeed, should j accept if $z_j > y_j(S, \delta)$ in this case? Our description is deliberately silent on these matters.

I will return to a formal argument later but for now, note that if player 1 is called upon to propose when all four players are present he will always make an unacceptable offer, provided that discount factors are close enough to 1.

The intuition is simply this: player 1 is a "weak partner" and therefore his proposals will be turned down by the other players unless he makes one of them — say player 2 — an offer equal to his "outside option", given by what 2 can get in her dealings with 3 or 4. In fact, it may be better for player 1 to simply let one of these other partnerships form and *then* proceed to deal with the remaining player on more equal terms.

The point of this exercise is not that you should be taking the bargaining delay, or even the making of unacceptable offers, seriously. The delay is in part an artifact of the assumption that player 1 *must* make a proposal and is not allowed to simply pass the initiative to another player, or to make a proposal on behalf of some other coalition. In both these cases the "delay" will go away; no unacceptable offers need be made. But the point is that an inequality like (7.2) will still hold strictly, and then the algorithmic characterization of equilibrium that we are about to attempt will no longer be possible. More on this below.

The alert reader will have noticed that Example 7.2 bears some similarities to Example 5.1 in Chapter 5, where a similar delay was noted in equilibrium. But that delay came from the presence of intercoalitional *externalities*, whereas at the root of this example lies *heterogeneity*. The possibility of a situation like Example 5.1 is ruled out by the Nondecreasing Average Worth (NAW) condition. The corresponding condition with heterogeneous players — even for characteristic function games — is more complex, and it's also different.[4]

We will presently move on to "no-delay" equilibria in which only acceptable proposals are made. But let us close this initial discussion with a useful property of equilibrium response vectors.

OBSERVATION 7.3. *Let* $\mathbf{y}(S, \delta)$ *be any equilibrium response vector, and suppose that (7.2) holds with equality for some $i \in S$. Then for any T that attains the maximum in (7.2) and for all $j \in T$, $y_j(S, \delta) \geq y_i(S, \delta)$.*

---

[4]Indeed, in a symmetric characteristic function, the condition that we will impose below will automatically be satisfied.

This is an important observation. It states that if a player makes an acceptable proposal in equilibrium, he either "proposes to himself"; i.e., chooses to stand alone, *or makes offers only to individuals who have at least as high an equilibrium response threshold as he does.*

While we relegate a formal proof to Section 7.8, it isn't hard to see why the observation is true. If i includes j in his proposal, j *could* include i in hers, and in so doing can pick up the same return i can. If j does indeed include i, it follows that $y_j = y_i$, and if she does not, this must be because she can do even better.

7.3 No-Delay Equilibrium

Say that an equilibrium is *no-delay* if after *every* history, every proposer makes an acceptable proposal. Then by Observation 7.2, the inequality (7.2) holds with equality for such equilibria.

7.3.1 Response Vectors for No-Delay Equilibrium. Let $\mathbf{m}(S, \delta)$ be the solution to the equality version of (7.2).

PROPOSITION 7.1. *For every* (S, δ), $\mathbf{m}(S, \delta)$ *exists and is unique.*

The existence of $\mathbf{m}$ relies on a perfectly standard argument based on Brouwer's fixed point theorem and I omit it. However, the asserted *uniqueness* of $\mathbf{m}$ is central to our analysis and it is worth putting a formal proof in the main discussion. As we shall see, the argument depends crucially on Observation 7.3.

Suppose, contrary to the claim, that there are two solutions $\mathbf{m}$ and $\mathbf{m}'$ to the full equality version of (7.2). Define K to be the set of all indices in S in which the two solutions differ; i.e., $K \equiv \{i \in S | m_i \neq m_i'\}$ and pick an index k such that one of these m-values is maximal. Without loss of generality suppose that it is the unprimed value m_k.

By definition, $m_k > m_k'$. Choose $T \subseteq S$ with $k \in T$ such that

(7.3)
$$m_k = \delta \left[v(T) - \sum_{j \in T-k} m_j \right].$$

Of course,

(7.4)
$$m_k' \geq \delta \left[v(T) - \sum_{j \in T-k} m_j' \right].$$

By Observation 7.3, $m_j \geq m_k$ for all $j \in T$. So, given our choice of $k \in K$, it must be that $m'_j \leq m_j$ for all $j \in T$. But then (7.3) and (7.4) together imply that $m'_k \geq m_k$, which is a contradiction.   □

The characterization of no-delay equilibria can be applied at once to deduce that there must be "delay" in Example 7.2 above. Simply calculate the **m**-vectors. For any two-player set with worth x, it is easy to see that

$$m_i(S, \delta) = \frac{\delta x}{1 + \delta} \text{ for } i \in S,$$

while for the grand coalition N, it is easy to check that

$$m_2(N, \delta) = m_3(N, \delta) = m_4(N, \delta) = \frac{100\delta}{1 + \delta}$$

$$m_1(N, \delta) = \delta \left[50 - \frac{100\delta}{1 + \delta} \right].$$

(You can verify all this by simply observing that the equality version of (7.2) is satisfied; after all, we already know that the solution is unique.)

Now $m_1(N, \delta)$, while always positive, converges to 0 as $\delta \rightarrow 1$. This proves that the equilibrium must exhibit delay for some initial choice of proposer. For if it didn't, player 1 can always make an unacceptable proposal. Then use the assumed continuation along the no-delay equilibrium to conclude that she can get a payoff bounded away from 0 (as $\delta \rightarrow 1$). This is a contradiction.

7.3.2 Condition M. Example 7.2, while exhibiting delay, implicitly suggests a condition that is sufficient for the existence of no-delay equilibrium:

[M] If $S \supseteq S'$, then for all $i \in S'$, $m_i(S, \delta) \geq \delta m_i(S', \delta)$.

Condition M is vaguely reminiscent of superadditivity, stating that players have access to larger m-values when the set of active players is larger, but it neither implies nor is implied by superadditivity.[5] Relative to the material covered in the last chapter, it is also an

---

[5]To show that [M] does not imply superadditivity, consider a two-player characteristic function in which $v(N) = 0$, while $v(i) = 1$ for $i = 1, 2$. Then [M] is satisfied. Conversely, consider the superadditive cover of the game in Example 7.2. It does not satisfy [M].

entirely new condition, because it is automatically satisfied in all symmetric games, and so does not make an appearance there.

These are somewhat cryptic and unsatisfactory remarks which we shall presently expand upon, but our first line of business is to see what Condition M implies:

PROPOSITION 7.2. *Under* [M], *every equilibrium must be no-delay.*

An equilibrium is described as follows. At any set of active players S a proposer i chooses some probability distribution over those coalitions T that maximize

$$v(T) - \sum_{j \in T-k} m_j(S, \delta)$$

and promises to deliver $m_j(S, \delta)$ to each $j \in T$. Any responder j accepts if and only if $y_k \geq m_k(S, \delta)$ for k equal to j and all those after him.

While a formal proof of the proposition is postponed to Section 7.8, it is easy enough to understand how it works. Why might delay occur in equilibrium? Example 7.2 explains. In that example, a proposer wishes to get others out of the way before she concludes her own dealings, so she stonewalls by making unacceptable offers. The reason she wants others out of the way is that she is in a stronger position once they are gone. While not entirely obvious, an inductive argument reveals that Condition M *rules out such a possibility.* Accordingly, under that condition, everyone wants to make an acceptable offer. But then (7.2) always holds with equality, and the equilibrium response vector must equal **m**. The proposition should now be clear.

7.4 Condition M, Payoffs and Coalition Structure

Notice that for every δ and every set of active players S, $\mathbf{m}(S, \delta)$ is a fixed point of an appropriate set of equations. This makes Condition M that much more cumbersome to check, though well-known numerical methods can be applied. More significantly, though, a "fixed-point description" of **m** makes it that much harder to understand Condition M, for we lack an intuitive description of the vector **m** itself.

The purpose of this section is to provide an algorithmic description of **m**, so that equilibrium responses are amenable to ready computation, and a restriction such as Condition M is simple to verify. This algorithm, while related to the one in the previous chapter, is distinct in two ways. First, we will set out the new algorithm for every discount factor, not just at the limit. This isn't a deep difference: with some more work, we could have done the same earlier, and indeed, we will pass to the no-discounting limit later in this chapter. More significantly, the algorithm we shall now describe comes into its own for asymmetric, heterogeneous situations. For the symmetric games studied in the previous chapter, characteristic functions posed no problem at all: all the hard work in that algorithm came from the presence of externalities. In contrast, what follows is nontrivial even in the externality-free world of characteristic functions.

7.4.1 An Algorithm. Fix a set of active players S and a discount factor δ. For any coalition $(C, S, \dots)$, use corresponding lower-case notation $(c, s, \dots)$ to denote cardinality.

Step 1. Begin by maximizing

$$\frac{v(C)}{1 + \delta(c - 1)}$$

over all $C \subseteq S$. Let C_1 be the *union* of all maximizers of this expression, and let a_1 be the maximum value attained. Define $m_i^* \equiv \delta a_1$ for every $i \in C_1$.

Step K+1. Recursively, suppose that we have defined sets $\{C_1, \dots, C_K\}$ for some index $K \geq 1$, corresponding values $\{a_1, \dots, a_K\}$, and have defined $m_i^* = \delta a_k$ whenever $i \in C_k$. Now define D to be the union of all the C_k's, and consider the following problem:

Choose *nonempty* $C \subset S - D$ and $T \subseteq D$ (but possibly empty) to maximize

$$\frac{v(C \cup T) - \sum_{i \in T} m_i^*}{1 + \delta(c - 1)}.$$

Define C_{K+1} to be the union of all sets C such that (C, T) maximizes this expression for some $T \subseteq D$, and let a_{K+1} be the maximum value attained. Define $m_i^* \equiv \delta a_{K+1}$ for every $i \in C_{K+1}$.

Continue in this way until **m** is fully defined on the set S of all active players.

The next proposition assures us that we've found the correct vector:

PROPOSITION 7.3. *For every set of active players S and discount factor δ, the vector* **m** *as constructed in the algorithm is precisely the collection* **m**$(S, δ)$ *of equilibrium response thresholds in no-delay equilibrium.*

This is a significant result in that it puts the direct computation of equilibrium payoffs for arbitrary characteristic functions within reach, provided of course that Condition M is satisfied.

A formal proof of this proposition may be found in Section 7.8, but a loose intuitive argument may be useful. We will need to show that for every individual $i \in S$,

$$m_i = \delta \left[v(W) - \sum_{j \in W-i} m_j \right].$$

for some coalition W containing i, and

$$m_i \geq \delta \left[v(W) - \sum_{j \in W-i} m_j \right].$$

for *every* coalition W containing i. By the uniqueness result of Proposition 7.1, that will be all that we need to do.

The first of these two requirements is easy enough to satisfy. If i belongs to the set C_{K+1} in the algorithm, simply choose the coalition that solves the maximization problem in Step $K + 1$. The second requirement will follow from the fact that at each step in the algorithm, a maximization problem is solved, so that non-maximizing coalitions generate a "discounted surplus" to a potential proposer i that is smaller than m_i. These steps form the basis of a formal proof.

We end this subsection by making an obvious but important remark. Algorithm in hand, it is now very easy to verify Condition M for any discount factor. Provided that that condition is met, Proposition 7.2 represents a powerful characterization of equilibrium. The next section explains how to make the link to equilibrium payoffs and coalition structure.

7.4.2 Equilibrium Payoffs and Coalition Structure. Assuming [M] holds, how is the algorithm applied to generate a set of equilibrium payoffs and associated coalition structure? Begin with the grand coalition N and suppose that player i proposes first. To use the algorithm, we first need to place i within the algorithm with active player set N: say she is located in the set C_K. Then apply Step K of the algorithm. She will make a proposal to one of the coalitions W of the form $C \cup T$ that solve the maximization problem in Step K. That proposal will be accepted. Player i's coalitional compatriots will enjoy payoffs that are (approximately) described by the projection of $\mathbf{m}(N, \delta)$ on W. Player i earns a bit more — $m_i(N, \delta)/\delta$ to be precise — but in any case all these payoffs coincide with (the T-projection of) $\mathbf{m}(N, \delta)$ as $\delta \to 1$.

If W is the full set of remaining players the game is over. Otherwise consider the remaining set of active players, say $S = N - W$. Now exactly the same process repeats itself: a proposer from S is chosen and she makes an acceptable proposal, but *now we must invoke the algorithm with active player set S.* In particular, equilibrium payoffs are *not* given by the vector $\mathbf{m}(N, \delta)$, even when δ approaches 1. That vector generates equilibrium payoffs for the *initial* proposer and the coalition to which that individual makes her proposal, but to go further we must use the corresponding vector for subcoalitions.

Thus equilibrium payoffs are generated by progressively marching down the list of active player sets as acceptable proposals are made, and using the $\mathbf{m}$-vector for *those* sets to generate payoffs at each stage.

In tandem with these recursively generated payoffs, a coalition structure forms, and we now examine this in more detail. At a proposer stage in which i proposes under active player set S, the coalition that forms is determined in the following way:

Let i belong to Step K of the algorithm in Section 7.4.1. Choose one of the coalitions that solve the maximization problem in Step K of that algorithm.

Now, it is entirely possible that there may be more than one coalition that solves the maximization problem in Step K. In that event — and unlike the symmetric case studied earlier — we will not have a precise prediction of equilibrium coalition structure even conditioning on proposer identity. What is more, this indeterminacy not only affects coalition structure but possibly equilibrium payoffs at subsequent stages of the game. Of course, this isn't to say that anything can happen — far from it — but the prediction isn't entirely tight either.

Such imprecision may simply be the price we have to pay for dealing with a very general class of characteristic functions. Fortunately, under a mild additional assumption, we can say quite a bit more:

[G.1] If S and T are two coalitions and at least one of them is not a singleton, then $v(S) \neq v(T)$.

Condition G.1 asks that characteristic functions come from some "generic" class. It takes heterogeneity seriously, arguing that no two worths should be exactly alike. Notice that the condition is not satisfied for symmetric games! As always, assumptions such as [G.1] or symmetry must always be applied with care, depending on the essential features of the application at hand.

That said, both symmetry and [G.1] yield strong uniqueness properties. We've already seen this for the symmetric case. For the model at hand, we have

PROPOSITION 7.4. *Under* [G.1], *there exists a discount factor* δ^* *such that if* $\delta \in (\delta^*, 1)$, *then the algorithm selects a unique coalition at every step, and this selection is entirely independent of the value of* $\delta \in (\delta^*, 1)$.

In particular, if Condition M holds for all discount factors close enough to 1, the same property is true of equilibrium: every proposer selects a unique coalition at every step, and these choices are independent of $\delta \in (\delta^*, 1)$.

While a formal proof is postponed to Section 7.8, it isn't hard to see how [G.1] enters the picture. For "generic" characteristic functions (as described by [G.1]), the presence of some discounting permits an essentially unique resolution to our algorithmic maximization problems, at least over a range for the discount parameter.

This is a good point to take stock of the progress so far, and to describe what comes next. We've concentrated on no-delay equilibria., those in which all proposers make acceptable offers at every stage. Such equilibria are very simply characterized by a collection of response vectors $\mathbf{m}(S, \delta)$, one for each set of active players S. Not only is this response vector uniquely defined, it can be fully described by a recursive algorithm. That algorithm, as we've just shown, also generates a unique coalition structure for every proposer protocol, provided that the underlying game satisfies a genericity condition.

Moreover, under Condition M, which guarantees that the response vectors are pointwise nondecreasing as the set of active players expands, every Markovian equilibrium must indeed be no-delay.

This is an essentially complete theory for no-delay equilibrium under very general conditions. However, it is also useful to consider an alternative route — one that we took in Chapter 5. Under this approach we develop an algorithm, as well as an analogue of Condition M, that depends on the parameters of the characteristic function *alone*, and can then be applied to all discount factors sufficiently close to one. Let us study this limit theory more carefully.

7.4.3 Low Bargaining Frictions. A limit computation has several advantages; we mention three. First, as we shall see below, it yields a solution concept that can be compared with other solutions from cooperative game theory. Second, it provides a quick and easily computable approximation which is valid when bargaining frictions are low. Finally, restrictions imposed directly on the limit (such as uniqueness) easily carry over to the case of low bargaining frictions.

The first (and extremely straightforward) point to note is that the algorithm holds without any change in the limit: simply set δ equal to 1 and run exactly the same recursive procedure. We summarize the main points in the following proposition:

PROPOSITION 7.5. *For every set of active players S, there exists a unique vector $\mathbf{m}^*(S)$ such that $\mathbf{m}(S, \delta)$ converges to $\mathbf{m}^*(S)$ as $\delta \to 1$.*

This vector can be computed by a limiting version of the algorithm in Section 7.4.1:

STEP 1. *Maximize $v(C)/c$ over $C \subseteq S$. Let C_1 be the union of all maximizers and let a_1 be the maximum value attained. Define $m_i^* \equiv a_1$ for every $i \in C_1$.*

Step $K + 1$. Recursively, suppose sets $\{C_1, \ldots, C_K\}$ are defined for some $K \geq 1$, and so are numbers m_i^* for every $i \in D = \cup_{k=1}^{K} C_k$. Choose (nonempty) $C \subset S - D$ and $T \subseteq D$ to maximize

$$\frac{v(C \cup T) - \sum_{i \in T} m_i^*}{c}.$$

Define C_{K+1} to be the union of all sets C such that (C, T) maximizes this expression for some $T \subseteq D$, and let a_{K+1} be the maximum value attained. Define $m_i^* \equiv a_{K+1}$ for every $i \in C_{K+1}$.

Continue in this way until $\mathbf{m}^*$ *is fully defined on the set S of all active players.*

Proposition 7.5 requires no proof, as it is simply a matter of passing to the limit in the algorithm of Section 7.4.1. But it is useful nonetheless. For instance, I could simply demand that the limit algorithm admit a unique coalitional solution at every step (see Condition U below). If that requirement were to be satisfied by some characteristic function, then there is no need to invoke additional genericity restrictions (such as [G.1]): the same uniqueness must automatically hold for all discount factors sufficiently close to 1.[6] We return to this approach in Section 7.6.5 below, when we study games with externalities.

The checking of Condition M is more subtle, however. We now have limit response vectors $\mathbf{m}^*(S)$, one for every set S of active players. We could simply ask that

$$(7.5) \qquad\qquad m_i^*(S) \geq m_i^*(S')$$

for every $S \supset S'$ and $i \in S'$. But this isn't enough to check Condition M for discount factors less than 1 or even sufficiently close to 1, and examples to support this claim are not hard to find. Certainly, one easy way around this problem is to hope that the stronger condition

$$(7.6) \qquad\qquad m_i^*(S) > m_i^*(S')$$

is satisfied for every $S \supset S'$ and $i \in S'$. But this is too restrictive a condition and excludes many interesting and relevant situations. The reason is that the relevant set for player i that determines her value of m_i (and the relevant sets for players in that set, and so on) may simply have nothing to do with players in $S - S'$, and in that case we would have equality: $m_i^*(S) = m_i^*(S')$. Loosely speaking, player i is "unaffected" by the change in the active player set.

To implement a working analogue of Condition M at the limit, then, we have to find a compromise between the weak restriction in (7.5) and the stronger restriction in (7.6). This is why the condition below is somewhat convoluted: the unaffected players need to be excluded.

In what follows, we assume

---

[6]It should be noted that Condition G.1 is not, in general, sufficient for uniqueness at the limit, so the two approaches are complementary.

[U] For every set of active players, the limit algorithm admits a unique coalitional solution at every step.

This means, in particular, that for every active set, each member is associated with a unique maximizing coalition.

(Recall from our earlier discussion that [U] guarantees the corresponding uniqueness of coalitional choice for all discount factors sufficiently close to 1.)

Now take two sets of active players S and S', with $S \supseteq S'$. Apply the algorithm in Proposition 7.5 for the smaller active set S'; it sequentially generates collections $\{C_1, \ldots, C_M\}$ of players and associated maximizing coalitions (uniquely defined, by [U]) $\{C_1, C_2 \cup D_2, \ldots, C_M \cup D_M\}$, where for each $k > 1$, $D_k \subseteq \cup_{j=1}^{k-1} C_j$.

If we were to carry out the corresponding algorithm for the larger active player set S, we would get a similar collection of player sets and associated maximizing coalitions. For some of the members of S', having the larger set S around may leave them completely unaffected: they are still associated with the same maximizing coalition and no other member of this coalition is affected either. In that case we must have $m_i^*(S) = m_i^*(S')$ for all "unaffected" i. We will place our condition on the remaining, "affected" individuals so we need to define this group more precisely. Say first that individuals in C_1 are *unaffected* if under the larger active set S, they continue to all be associated with the same maximizing coalition C_1. Recursively, having defined the concept for all $j = 1, \ldots, k-1$, say that individuals in C_k are *unaffected* if under the larger active set S, they continue to all be associated with the same maximizing coalition $C_k \cup D_k$, and if all members of D_k are unaffected. Finally, say that an individual is *affected* if she is not unaffected.

Now for the new version of Condition M:

[M*] If $S \supset S'$, then for all affected $i \in S'$, $m_i^*(S) > m_i^*(S')$.

This is a variant that can be directly checked at the limit, and it will assure us that [M] indeed holds for all discount factors large enough.[7]

We summarize our discussion in

---

[7]It is also possible to verify that [M*] is generic, relative to the universe of characteristic functions in which Condition M is satisfied.

PROPOSITION 7.6. *Impose the generic restriction that for every set of active players, the limit algorithm admits a unique coalitional solution at every step. Suppose, moreover, that Condition M* holds.*

Then there exists a discount factor δ^* such that if $\delta \in (\delta^*, 1)$, the equilibrium is no-delay. It is described as follows: at any set of active players S a proposer makes an acceptable proposal to the unique coalition that contains him in the limit algorithm for the set S.

The payoff for each member i in that coalition converges to $m_i^*(S)$ as $\delta \to 1$.

The discussion leading up to this proposition constitutes an informal proof of it; we fill in the details in Section 7.8.

7.4.4 What Does m* Look Like?.

It is extremely easy to compute the **m**-vector for any given characteristic function, and even easier to calculate the limit vector **m***. The most trivial example is

EXAMPLE 7.3. *Two-Person Bargaining.* $N = \{12\}$, $v(N) = 1$ *and* $v(i) = 0$ *for* $i = 1, 2$.

Then for singleton coalitions, $m_i(\{i\}, \delta) = 0$, while

$$m_1(N, \delta) = m_2(N, \delta) = \frac{\delta}{1 + \delta},$$

which converge to a limit vector of $1/2$ each as δ goes to 1. Obviously Condition M is satisfied, so Proposition 7.2 applies. Using the algorithm of Section 7.4.1, each proposer makes a proposal to the grand coalition, and share the resulting surplus almost equally as $\delta \to 1$.

Of course, all this is well-known and this example is not meant to uncover new insights. I put it up simply to benchmark the **m**-vector using an entirely familiar problem. Now here is a variant.

EXAMPLE 7.4. *Two-Person Bargaining With Outside Options.* $N = \{12\}$, $v(N) = 1$, $v(1) = a$, $v(2) = b$. *Assume* $1/2 \neq a > b \geq 0$ *and* $a + b < 1$.

I've put in the superadditivity restriction simply to cut down on the number of cases. It is easy enough to compute the limit vector **m***(N):

$$m_1^*(N) = \max\{a, 1/2\} \text{ and } m_2^*(N) = 1 - m_1^*(N).$$

One can easily check that condition M* holds, so that Proposition 7.6 is applicable. It's easy to see that we are back to the usual bargaining model if $a < 1/2$. Otherwise, if a exceeds $1/2$, player 1 obtains precisely this amount as proposer (for low bargaining frictions), whereas player 2 as proposer picks up the rest of the surplus. This is an interesting observation. Player 1's superior outside option acts as a constraint on the bargaining outcome, in the sense that she obviously cannot be pushed down below that outside option. But as bargaining frictions vanish, she is unable to do *better* than her outside option, if it amounts to more than half the bargaining surplus.

Some might find this an odd result. For instance, those familiar with Nash bargaining might insist that the the available surplus — over and above outside options — "should" be split equally, so that player 1 "should" get $a + (1 - a - b)/2$. But I've put the word "should" in quotes to deliberately suggest that such statements are entirely a matter of taste, or at least a matter of what one is used to. One could equally well argue that bargaining is a process which is inherently geared towards egalitarianism — towards equal division — and this tendency can only be halted by the (credible) threat of walking away from the process, which is provided by the outside option. So one might equally expect to see "the most equal division of worth subject to outside options".

In Section 7.7 I examine the robustness of this particular implication of our bargaining model (the discussion in Section 14.2, Chapter 14, is also relevant). But it is worth noting that the "constrained egalitarianism" displayed by our model is closely connected to a solution concept in cooperative game theory: the *egalitarian solution* developed by Bhaskar Dutta and myself in 1989.[8] Roughly speaking, the egalitarian solution searches for the Lorenz-maximal allocations of coalitional worth subject to participation constraints. Dutta and Ray (1989) prove that characteristic functions have a remarkable property: despite the fact that the Lorenz ordering is a partial one,[9] *there is at most one egalitarian solution for each coalition.* Using the limit algorithm developed here, it is possible to show that our limit $\mathbf{m}^*$ vector and the egalitarian solution are closely linked. Chatterjee, Dutta, Ray and Sengupta (1993) prove that the two are exactly the same for all *convex* games, those with the property that for all coalitions S and T, $v(S \cup T) \geq v(S) + v(T) - v(S \cap T)$. Considerations

---

[8]See Dutta and Ray (1989, 1991).

[9]It is equivalent to second-order stochastic dominance if allocations are viewed as wealth distributions.

of space preclude a more detailed development of this idea, though we briefly return to convex games in Section 7.5 below.

The tendency towards egalitarianism is sometimes responsible for the exclusion of certain coalitional members. In the example that follows, a coalition of players excludes others even though there are potential gains from including them.

EXAMPLE 7.5. *Exclusion.* $N = \{123\}$, $v(\{12\}) = 3$, $v(\{123\}) = 4$, *while* $v(S) = 0$ *for all other S.*

Applying the first step of our limit algorithm, it is easy to see that $v(S)/s$ is maximized at the coalition $\{12\}$. One can go the rest of the way and fully verify [M*]. It follows right away that players 1 and 2 will never propose the grand coalition when bargaining frictions are small. One interpretation is that players 1 and 2 fear the redistributive impact of allowing player 3 to join the coalition: a worth of 3 split two ways is better than a worth of 4 split three ways.

On the other hand, player 3 is desperate to form the grand coalition, as Step 2 of the algorithm will readily show. Given the credible tendency towards exclusion displayed by players 1 and 2, she is willing to pay them their higher outside options in return for a (smaller) share of the surplus.[10]

This example also shows that equilibrium coalition structure cannot be generally disassociated from the order of proposers. In non-symmetric games, different proposers will generally want to form different coalitions. There isn't any getting away from this fact, and a theory which attempts to do so would, in my opinion, be attempting sharper predictions at the expense of a severe loss of realism.

We will return to the theme of exclusion below. But it is worth noting an interesting property of the equilibrium outcome when [M] (or [M*]) holds and at least one player does *not* wish to exclude anyone else, which is the case with player 3 in Example 7.5. *Then the limit vector* $\mathbf{m}^*(N)$ *must lie in the core of the game.*

The reason is simple, and is related to a general property of the $\mathbf{m}^*$-vector. For every i and every coalition S containing i, it is always

---

[10]This feature is also shared by Example 7.4. If player 1 proposes first, she will simply walk away with a, while if player 2 proposes first, she will want to form the grand coalition.

true that

$$m_i(N, \delta) \geq \delta \left[v(S) - \sum_{j \neq i} m_j(N, \delta) \right]$$

so that if we pass to the limit as $\delta \to 1$, it must be that

(7.7)
$$\sum_{i=1}^{n} m_i^*(N) \geq v(S).$$

In particular, it's always the case that

(7.8)
$$\sum_{i=1}^{n} m_i^*(N) \geq v(N).$$

When the inequality in (7.8) fails to hold strictly — and it must fail when one player behaves in a non-exclusionary way — the vector $\mathbf{m}^*(N)$ is a core allocation.

My final example returns to the theme of exclusion. Unlike the previous examples, there are situations in which the final outcome involves exclusion, irrespective of who proposes.

EXAMPLE 7.6. *The Employer–employee Game.* $N = \{123\}$, $v(i) = 0$ *for all* i, $v(\{23\}) = 0$, $v(\{12\}) = v(\{13\}) = 1$, $v(N) = 1 + \mu$ *for some* $\mu > 0$. *The interpretation is that player 1 is an employer who can produce an output of 1 with any one of the two employees 2 and 3. He can also hire both employees in which case output is higher. No other combination can produce anything.*

It is very easy to compute the **m**-vector. For two-person coalitions,

$$m_i(\{12\}, \delta) = m_j(\{13\}, \delta) = \frac{\delta}{1 + \delta},$$

for $i = 1, 2$ and $j = 1, 3$, while

$$m_k(\{23\}, \delta) = 0$$

for $k = 2, 3$. For the grand coalition, an application of the algorithm reveals that *for values of δ sufficiently close to 1,*

$$m_i(N, \delta) = \frac{\delta}{1 + \delta}$$

for all i, provided that $\mu < 1/2$, while

$$m_i(N, \delta) = \frac{\delta(1 + \mu)}{1 + 2\delta}$$

for all i, when $\mu \geq 1/2$. In either case one can check that [M] holds.

The algorithm can now be used to predict equilibrium coalition structure for δ close to 1. If $\mu \geq 1/2$, then each player proposes to the grand coalition (the exact terms are described in Proposition 7.2) and the proposal is accepted, yielding a three-way equal division of $1 + \mu$ as $\delta \to 1$. If $\mu < 1/2$, then each player will make an acceptable proposal to a *two*-player coalition, so that either coalition {12} or coalition {13} will form (dividing their worth equally in the limit), leaving the third player out.

Some features of this example, such as approximate equal division among the coalitions that do form, are not robust to alternative bargaining protocols. What *is* robust, though (and we shall see this in Section 7.7), is that some degree of inefficiency is endemic. Indeed, under the particular protocol we consider, *the equilibrium outcome is inefficient no matter who proposes first.* I take up this theme in more detail in the next section.

7.5 More on Efficiency

We exploit the characterization in Proposition 7.2 to throw light on the question of efficiency. As I have made clear at several points, the bargaining process contains an implicit externality: when someone makes an offer, she has to adequately compensate her responders. Otherwise they can seize the initiative. This means that "at the margin" when a proposer is choosing a coalition, part of the surplus from that coalition has to be "given away". In this way a wedge is driven between the "private surplus" and the "social surplus", which often results in an inefficient choice by the proposer.

To see this from another angle, consider the dictator version of our game in which only one player gets to make offers and everyone else can only say yes or no. In that case the outcome is efficient in all equilibria because the entire social surplus is appropriated by the dictator who therefore maximizes that surplus. The outcome may not be very welcome from an equity point of view but that is another matter. However, once we depart from the dictator version by allowing a responder the power to formulate her own counterproposal, the wedge is resurrected.

It is possible to distinguish between two notions of efficiency. In the first notion, *every* initial proposer generates an efficient outcome. Call this *strong efficiency*. In contrast, define *weak efficiency* to be a property of equilibrium in which at least one initial proposer generates an efficient outcome. What follows are some observations on these different notions of efficiency.

First, it is possible to exactly describe those games that permit strong efficiency, at least for low bargaining frictions.

PROPOSITION 7.7. *A coalitional bargaining game with a strictly superadditive characteristic function is strongly efficient for all discount factors close to 1 if and only if*

(7.9) $$\frac{v(N)}{|N|} \geq \frac{v(S)}{|S|} \text{ for all coalitions } S.$$

The proposition tells us that strong efficiency is possible, but only under fairly demanding circumstances. The grand coalition must have at least as high an average worth than any other coalition. If we assume Condition M (which the proposition does not do), we then have the algorithm in place, and it isn't hard to get a sense of why the proposition must be true. Suppose that bargaining frictions are low. If the equilibrium outcome is to be efficient no matter who proposes, the algorithm must terminate in a single step, with the grand coalition a unique maximizer of $v(C)/c$ over all coalitions C, in that step. So (7.9) must hold. The formal proof in Section 7.8 handles the general case in which [M] is not assumed.

Notice that condition (7.9) is precisely the condition for a nonempty core in symmetric games (see Observation 7.1). Of course, that condition, while also sufficient for a nonempty core in general asymmetric games, is not necessary. Yet the connection can hardly be missed: the proposer-independent efficiency of a game is intimately tied to a condition that is even stronger than the nonemptiness of the core. Let us examine this relationship a bit more carefully.

To do so, it will be useful to consider the milder notion of weak efficiency: for *some* initial proposer, the outcome is efficient. It turns out that weak efficiency reveals a more direct connection to a nonempty core (rather than to a sufficient condition for the core to be nonempty).

PROPOSITION 7.8. *Suppose that (N, v) has a strictly superadditive characteristic function and suppose that we have weak efficiency along a sequence of equilibria, under some sequence of discount factors converging to one. Let $\mathbf{z}(\delta)$ be the corresponding sequence of efficient equilibrium payoff vectors. Then any limit point of $\mathbf{z}(\delta)$ lies in the core of (N, v).*

The proof of this proposition is simple and intuitive and worth recording as part of the main discussion. Suppose, then, without loss of generality, that player 1 is the "efficient proposer" along some sequence of weakly efficient equilibria.[11] By efficiency and strict superadditivity, she must be making an acceptable proposal to the grand coalition. Then, if $y(N, \delta)$ is the equilibrium response vector, we have

(7.10)
$$z_1(\delta) = \frac{y_1(N, \delta)}{\delta},$$

while

(7.11)
$$z_j(\delta) = y_j(N, \delta)$$

for all other j. Now pick any coalition S and any $i \in S$. By (7.2),

$$y_i(N, \delta) \geq \delta \left[v(S) - \sum_{j \in S - i} y_j(N, \delta) \right],$$

or

$$\frac{y_i(N, \delta)}{\delta} + \sum_{j \in S - i} y_j(N, \delta) \geq v(S).$$

Pick a subsequence such that $z(\delta)$ converges, say to z^*. Send δ to 1 along this subsequence and note that both $y_i(N, \delta)$ and $y_i(N, \delta)/\delta$ converge to z_i^*. Therefore

$$\sum_{i \in S} z_i^* \geq v(S).$$

Because S was arbitrarily chosen, we are done.   □

Now the connection with the core starts to become clearer. For discount factors close to 1, games with empty cores will *never* have efficient stationary equilibria no matter who proposes first. That is, even weak efficiency is not possible.

What about the converse? If equilibria are inefficient, must the core be empty? To see this, recall the employer–employee game from Example 7.6: $N = \{123\}$, $v(i) = 0$ for all i, $v(\{23\}) = 0$, $v(\{12\}) = v(\{13\}) = 1$, $v(N) = 1 + \mu$ for some $\mu > 0$. We know from our discussion of that example that if $\mu < 1/2$, inefficiency is endemic: the efficient grand coalition never forms no matter who proposes.

---

[11]Of course the efficient proposer may change along the sequence, but then simply take an appropriate subsequence.

Therefore even weak efficiency is not to be had. *Yet it is easy to verify that the core of this game is nonempty.*

Player 1 would love to include both players 2 and 3 and pocket the resulting surplus, but the very act of including both players gives them bargaining power (because they can reject an offer). Both players will have to be compensated at overall rates that do not justify the gain of the extra μ. (The rates would be justified if $\mu > 0.5$.)

Fortunately, the limit result of Proposition 7.5, as well as the subsequent discussion of $\mathbf{m}^*$, helps in establishing a sharper criterion for inefficiency, one that incorporates both the empty core as well as situations such as Example 7.6. Recall the property of the $\mathbf{m}^*$-vector described in (7.8):

$$(7.12) \qquad \sum_{i=1}^{n} m_i^*(N) \geq v(N).$$

This inequality is perfectly general. When it holds *strictly*, weak efficiency must fail for low bargaining frictions:

PROPOSITION 7.9. *For a strictly superadditive game satisfying Condition M at δ close to 1, recall the limit vector $\mathbf{m}^*$ described in Proposition 7.5. If $\sum_{i=1}^{n} m_i^*(N) > v(N)$, then no sequence of equilibria can be weakly efficient under any sequence of discount factors approaching 1.*

The proof of this proposition is very simple, so we omit a formal account. By [M], every player must make an acceptable offer, and obtain a payoff of $m_i^*(N, \delta)/\delta$ in her role as proposer. Because (7.12) holds with strict inequality, and because $\mathbf{m}(N, \delta)$ converges to $\mathbf{m}^*(N)$ as $\delta \to 1$, such a proposal *cannot* be made to the grand coalition when bargaining frictions are small. It follows from strict superadditivity that an efficient outcome isn't to be had.

The reason that Proposition 7.9 is a more expansive version of its predecessor (at least within the restrictions imposed by Condition M) is that core emptiness *implies* strict inequality in (7.12), which is the starting premise of Proposition 7.9. Recall the discussion in Section 7.4.4: if, on the contrary, $\sum_{i=1}^{n} m_i^*(N)$ equals $v(N)$, it must be a core allocation.

Proposition 7.9 also includes the employer–employee game from Example 7.6. There, the core is empty but it is easy to see that (7.12) holds strictly.

The *failure* of efficiency (weak or strong) is the notable feature of our analysis, and therefore our natural focus is on easily verifiable sufficient conditions for inefficiency. At the same time, the reader might also be interested in sufficient conditions for weak efficiency, and so I end this section with a couple of remarks on the subject.

In fact, it isn't easy to establish a satisfactory converse to Proposition 7.9. In the converse situation, (7.12) fails to hold strictly; that is,

$$(7.13) \qquad \sum_{i=1}^{n} m_i^*(N) = v(N).$$

It should be noted that this equality condition is *not* knife-edge. For example, every two-person superadditive game satisfies it. Moreover, it is easy to see that every superadditive two-person game is weakly efficient. However, (7.13) is not, in general, sufficient for weak efficiency. Consider the following example:

EXAMPLE 7.7. *Failure of Weak Efficiency Even When (7.13) Holds.* $N = \{123\}$, $v(1) = 1$, $v(2) = v(3) = 0$, $v(\{12\}) = 1.8$, $v(\{13\}) = 1.6$, $v(\{23\}) = 0.1$, *and* $v(\{123\}) = 2.4$.

This is a strictly superadditive game and Condition M holds for all discount factors close to one. Moreover, it is possible to check that for large discount factors,

$$\mathbf{m}(N, \delta) = (\delta, 1.8\delta - \delta^2, 1.6\delta - \delta^2),$$

so that (7.13) indeed holds. Yet no player proposes the grand coalition in equilibrium for discount factors close to 1. Weak efficiency fails.

To understand the example better, consider a strictly superadditive game satisfying Condition M. Recall the algorithm that defines $\mathbf{m}(N, \delta)$. Refer to an individual i with the lowest value of $m_i(N, \delta)$ as a *weakest player*. Notice that either some weakest player proposes the grand coalition, or *no one* does, and weak efficiency must fail. This assertion is an immediate corollary of Observation 7.3, which states that any player makes an acceptable offer only to those with an equilibrium response at least as high as hers.

With this in mind, notice that player 3 is the weakest player in Example 7.7, when bargaining frictions are small. So weak efficiency hangs on her making an offer to the grand coalition. Her payoff from doing so is

$$2.4 - \delta - [1.8\delta - \delta^2],$$

but her payoff from forming the smaller coalition {13} is $1.6 - \delta$, which is easily seen to exceed the former expression for all discount factors close enough to 1. This is enough to destroy weak efficiency altogether.

Thus weak efficiency is intimately related to the willingness of the weakest player to include everyone else in his proposed coalition. Chatterjee, Dutta, Ray and Sengupta (1993) prove that such a willingness can always be demonstrated when the game in question is strictly convex. For completeness, we state this proposition but omit the proof; the reader interested in a more detailed treatment may consult the cited paper.

PROPOSITION 7.10. *Suppose that a characteristic function game is strictly convex, so that $v(S \cup T) > v(S) + v(T) - v(S \cap T)$ for all coalitions S and T. Then for all discount factors close enough to one, Condition M is automatically satisfied, and the game is weakly efficient.*

As a closing remark, notice that all strictly superadditive two-person games are also strictly convex, and so the weak efficiency of such games (at least for low bargaining frictions) is an immediate corollary of Proposition 7.10.

As a summary, then, efficiency is far from guaranteed in characteristic function games. Superadditive games with empty cores are invariably inefficient in a strong sense: for no proposer is the outcome efficient. But the condition for this sort of efficiency failure is actually weaker: there are games with nonempty cores in which equilibrium outcomes are inefficient (Proposition 7.9 illustrates).

7.6 Externalities Revisited

Now it is time to take our analysis a (large) step further, by bringing intercoalitional externalities back into the picture. This additional layer of complexity builds on the analysis conducted so far, and the serious reader interested in a full understanding should be reasonably comfortable with the preceding discussion before proceeding further.[12]

---

[12]Relative to the symmetric case, even the arguments made for characteristic functions are quite complex. This isn't surprising. After all, we're attempting to formulate an approach that will work for a quite general class of characteristic function games.

The primitive of the analysis is now a *partition function v*, which assigns to every coalition structure π and every coalition S in that coalition structure a worth $v(S, \pi)$. (We continue to impose the restriction that payoffs are transferable.) Given our interest in asymmetric situations, there is no longer any presumption that coalitions of the same size generate the same worth. Indeed, the following "genericity condition" squarely removes us from the symmetric world:

[G.2] If S and T are two distinct coalitions (at least one of which is not a singleton), then $v(S, \pi) \neq v(T, \pi')$ for any pair of coalition structures π and π' such that $S \in \pi$ and $T \in \pi'$.

Obviously, [G.2] extends our earlier condition G.1 to partition functions. As in the case of [G.1], we've been a bit heavy-handed in our choice of assumption; something much weaker would have sufficed. But the extra details involved in the exposition are probably not worth the gain in generality.[13]

We need some more discussion of the protocol. As we have been doing in this chapter, we continue to restrict ourselves to the rejector-proposes protocol (though see Section 7.7 below). But we will also need to be more explicit about what happens after an accepted proposal. How is the new proposer chosen for the remaining set of active players? Thus far we did not need to worry about the details of this choice: for symmetric games, all proposers generated the same numerical coalition structure (which is all that mattered), and for characteristic function games the identity of the next proposer was irrelevant anyway (because there were no externalities). When the situation is asymmetric *and* has externalities, we need to be more specific.

We will suppose that for every active set of players an initial proposer is chosen according to some deterministic rule, and thereafter the rejector-proposes protocol takes over.

I should add right away that the analysis could just as readily be conducted with any rule, deterministic or stochastic, that chooses the initial proposer from a set of active players. The approach that we take will continue to hold for a generic class of games.[14]

---

[13]Considerable headway can be made by imposing the restriction in [G.2] on S and T only when they have the same cardinality.

[14]Briefly, given the rule, the genericity assumption [G.2] will need to be rephrased relative to that rule. It will still be of the form "the worth of S is different

7.6.1 No-Delay Equilibrium. Our goal, as before, is to describe no-delay equilibria, those in which every proposer in every subgame makes an acceptable proposal. The first main result of this section is

PROPOSITION 7.11. *Under [G.2], there exists a discount threshold δ^* such that for every $\delta \in (\delta^*, 1)$, at most one no-delay equilibrium exists. The equilibrium coalition structure will depend on the order of proposers, but the precise choice of coalition and allocation is fully determined for every proposer at every stage of the game. Moreover, the choice of coalition is independent of $\delta \in (\delta^*, 1)$.*

The proof of this central proposition essentially involves the piecing-together of various results that we have already established. In what follows we do just this. We break up the analysis into distinct steps.

7.6.2 Coalition Maps and Completion Maps. First, we introduce the idea of a "coalition map". Let π° be the collection of all coalition *sub*structures; i.e., the collection of all coalition structures on every *strict* subset of N. (The "empty structure" ϕ is included in this collection.) For every substructure $\pi \in \pi^\circ$, denote by $S(\pi)$ the *remaining* set of active players.

A *coalition map* Λ assigns to every substructure $\pi \in \pi^\circ$ a fresh coalition $\Lambda(\pi)$ drawn from the set $S(\pi)$.

Coalition maps are the obvious extension of the algorithmic mapping $t(\mathbf{n})$ in the symmetric case, which assigned a fresh *numerical* coalition to every *numerical* substructure that has "already formed". There isn't anything different here; it's just that we now need to keep track of coalitions rather than coalitional sizes.

Given any coalition map Λ, a substructure can be "completed" into a full coalition structure of the entire player set in the obvious way. Simply apply the map recursively starting from the substructure in question, until no players are left. In this way, I can generate a new map which maps substructures to "completed" coalition structures; call this the *completion map* $c(\cdot, \Lambda)$ associated with the coalition map Λ. It will be notationally useful to define $c(\pi, \Lambda) \equiv \pi$ for all (full) coalition structures π.

---

from that of T whenever $S \neq T$," but these worths will have to be calculated using probability mixtures over coalition structures in proportions related to the distribution governing a choice of initial proposer.

7.6.3 Characteristic Functions From Coalition Maps. For every coalition map Λ, a corresponding characteristic function $v_{\Lambda\pi}$ on the remaining player set $S(\pi)$ is induced. This function is defined as follows. For any coalition T from the player set $S(\pi)$, apply the completion map to the structure $\pi.T$ to obtain a full coalition structure $c(\pi.T, \Lambda)$. Record the worth of T under this structure:

$$(7.14) \qquad v_{\Lambda\pi}(T) \equiv v(T; c(\pi.T, \Lambda)).$$

Repeat the process for every such coalition. A characteristic function has now been obtained.

Because we've assumed the genericity condition G.2, every such $v_{\Lambda\pi}$ must satisfy the genericity condition G.1 for characteristic functions: $v_{\Lambda\pi}(S) \neq v_{\Lambda\pi}(T)$ whenever S and T are distinct (and at least one of them is nonsingleton).

It follows right away that Proposition 7.4 must apply to each such characteristic function: there exists a threshold discount factor such that for all discount factors exceeding this threshold, every no-delay equilibrium must single out the same, unique coalitional choice for every proposer. The threshold conceivably depends on both Λ and π, but as there are only finitely many coalition maps and substructures, we can find one threshold discount factor δ^* that does the job for all of them.

7.6.4 Consistent Coalition Maps. The observations in the previous section point the way to a proof of Proposition 7.11.

Fix any discount factor greater than δ^*. Say that a coalition map Λ is *consistent* if for every substructure $\pi \in \pi^\circ$, $\Lambda(\pi)$ is exactly the same as the coalition uniquely chosen under the algorithm and containing the initial proposer at player set $S(\pi)$.

A consistent coalition map provides the appropriate generalization of the algorithm for the symmetric case. Under a consistent coalition map, a substructure will be "completed" exactly along equilibrium lines, and those equilibrium choices in turn will justified at every step of the way by the (correct) expectation that the coalition map in question will be faithfully applied at all future stages of the game. Every consistent coalition map must correspond to a no-delay equilibrium, and vice versa.

But more can be said: under the assumptions of Proposition 7.11, each protocol that deterministically assigns an initial proposer (for every set of active players) is associated with *one and only one* consistent coalition map, provided that $\delta > \delta^*$. Verifying this is a simple matter of backward induction. Begin with substructures π such that $S(\pi)$ is a singleton; uniqueness is trivial here. Recursively, suppose that the map Λ is uniquely defined for all substructures that nest a given substructure π. Then the completion $c(\pi.T, \cdot)$ is uniquely defined for every choice of coalition T at that substructure, and so is the corresponding characteristic function $v_{\Lambda\pi}$ (see (7.14)). Now the algorithm kicks in, and the uniqueness theorem in Proposition 7.4 assigns precisely one coalition at this stage. Therefore the coalition map at π can assume precisely one value at this stage, and the recursive step needed to prove Proposition 7.11 is complete.

7.6.5 More on Uniqueness. As discussed earlier for characteristic functions, there are other ways to obtain uniqueness in no-delay equilibrium. Returning for a moment to the special case with no externalities, recall the direct computation of the limit vector $\mathbf{m}^*$ (and its associated coalitions) described in Proposition 7.5. If that computation yields — by assumption — a unique coalition at every algorithmic step, we are done: for discount factors close enough to 1, the no-delay equilibrium coalitions chosen by each player must coincide with the coalitions described in that algorithm, and there will be only one coalition for each proposer.

This alternative approach to uniqueness can be pursued here as well. It may be worthwhile to record it explicitly. Once again, we will need to invoke a genericity condition that takes heterogeneity seriously. To develop the condition, fix a coalition map Λ. For any substructure π, define the associated characteristic function $v_{\Lambda\pi}$ just as we did in equation (7.14) of Section 7.6.3. The new restriction can now be stated as

[U*] For every coalition map Λ and substructure π, the algorithm used to compute the limit $\mathbf{m}^*$-vector for the associated characteristic function $v_{\Lambda\pi}$ admits a unique coalitional solution at any step.[15]

The condition looks strong, imposed as it is for every coalition map and substructure. Like its counterpart [U] for characteristic functions, however, it is a bona fide genericity condition: if a partition

---

[15]A weaker but more complicated version that permits closer comparison with [G.1] and [G.2] can be obtained by permitting multiple solutions *only if* each such solution is a singleton coalition. The same uniqueness result follows.

function were to be chosen from some continuous distribution over the space of all payoff vectors, [U*] would almost always be satisfied.

With [U*] in hand, the generation of an equilibrium coalition structure via the device of a consistent coalition map (see Section 7.6.4) must invariably result in a unique solution for each ordering of initial proposers. We therefore have the following proposition, which needs no proof but is worth recording formally:

PROPOSITION 7.12. *Under [U*], there exists a discount threshold δ^* such that for every $\delta \in (\delta^*, 1)$, at most one no-delay equilibrium exists. The equilibrium coalition structure will depend on the order of proposers, but the precise choice of coalition and allocation is fully determined for every proposer at every stage of the game. Moreover, for each proposer, the choice of coalition is independent of $\delta \in (\delta^*, 1)$.*

7.6.6 An Algorithm for the General Case. Just as in the case of characteristic functions, we can describe equilibrium coalition structure and payoffs by means of an algorithm. The algorithm combines the recursive method used for symmetric games (Section 5.2 in Chapter 5) with the limit computation introduced in Proposition 7.5, and incorporates within it the idea of a consistent coalition map.

Recall the space π° of coalition substructures (which includes the "null substructure" ϕ). Just as in the algorithm for symmetric games, we are going to construct a (consistent) coalition map $\Lambda(\pi)$ that assigns to each member π of π° a coalition drawn from the remaining set of players $S(\pi)$.

By applying this coalition map repeatedly starting from ϕ, we will generate a particular numerical coalition structure, to be called π^*. We will also be interested in an associated vector of payoffs, to be described below.

Throughout, we impose the generic restriction [U].*

STEP 1. For all π such that $S(\pi)$ is a singleton, define $\Lambda(\pi) \equiv S(\pi)$.

STEP 2. Recursively, suppose that we have defined $\Lambda(\pi)$ for all substructures π with $|S(\pi)| \leq k - 1$ for some integer $k \geq 2$. For any such π, define the associated completion map $c(\pi, \Lambda)$, which is a full coalition structure.

STEP 3. For any substructure π such that $|S(\pi)| = k$, construct the characteristic function

$$v_{\Lambda\pi}(T) \equiv v(T; c(\pi.T, \Lambda))$$

for all nonempty $T \subseteq S(\pi)$, just as in equation (7.14).

STEP 4. Use the limit algorithm of Proposition 7.5 to obtain a vector $\mathbf{m}^*(\pi, \Lambda)$ and associated maximizing coalitions for the characteristic function $v_{\Lambda\pi}$, on the player set $S(\pi)$.

STEP 5. The protocol determines an initial proposer at the player set $S(\pi)$, say i. By [U*], there is a unique maximizing coalition from the limit algorithm in Step 4 that contains player i. Set $\Lambda(\pi)$ equal to this coalition.

STEP 6. Complete this recursive definition so that Λ is now defined on all of π°. Define a full coalition structure (for the entire set of players) by

$$\pi^* \equiv c(\phi, \Lambda).$$

STEP 7. For any coalition $S \in \pi^*$, let π_S^* stand for the substructure that immediately precedes it in π^*. For each $i \in S$, define $\widehat{m}_i \equiv m_i^*(\pi_S^*, \Lambda)$.

Note that the vector $\widehat{\mathbf{m}}$ is slightly different from the $\mathbf{m}^*$-vector that we used for characteristic functions. It is "built" from the $\mathbf{m}^*$-vector, however, and the given proposer protocol. It assigns the $\mathbf{m}^*(\phi, \Lambda)$-vector (or the appropriate projection of it to be precise) to the very first of the coalitions $\Lambda(\phi)$. It then draws values from the vector $\mathbf{m}^*(\Lambda(\phi), \Lambda)$ for the second coalition; and so on until all players have been accounted for. As the following proposition makes clear, $\widehat{\mathbf{m}}$ is a full description of the equilibrium payoff vector at the limit, just as π^* is a full description of coalition structure:

PROPOSITION 7.13. *Under [U*], and with a given initial proposer at every active player set, there exists a threshold $\delta^* \in (0,1)$ such that for all $\delta \in (\delta^*, 1)$, a no-delay equilibrium must yield the coalition structure π^* described in Step 6 of the algorithm. Moreover, the equilibrium payoff vector must converge (as $\delta \to 1$) to $\widehat{\mathbf{m}}$, described in Step 7 of the algorithm.*

Subject to the generic restriction [U*], Proposition 7.13 is the overarching theorem of this chapter. It has many pieces embedded in it, which have been built up in great detail over the course of the chapter. With those pieces in place, the proof of the proposition simply consists in joining all the fragments into one connected argument.

There remains one missing step. The proposition fully characterizes no-delay equilibrium. But we still need to assure ourselves that a no-delay equilibrium exists. This is the subject of the next section.

7.6.7 The Existence of No-Delay Equilibrium.
The argument so far *assumes* that equilibria are no-delay, and predicts equilibrium coalition structure on that presumption. We end our discussion by providing sufficient conditions under which this will indeed be the case.

It is useful to recall two sets of conditions that guaranteed the existence of no-delay equilibrium in special cases. The first is the nondecreasing average worth condition (NAW) that we invoked for *symmetric* games with externalities. The second is Condition M (or its limit counterpart [M*]), which guaranteed the existence of no-delay equilibrium in asymmetric games *without* externalities. It will come as no surprise that a similar sufficient condition for the general model must somehow combine NAW and condition M (or [M*]).

We will continue to assume [U*], and will therefore work with the limit algorithm from the previous section. That algorithm provides a vector of payoffs $\widehat{m}$ and a coalition structure π^* of the active players. By using exactly the same method, we could construct corresponding objects for *any* set of active players, after a substructure π "exits". In addition, one could also assume any arbitrary player i to be the first proposer and follow the given proposer protocol after that. The algorithm would just as easily generate a payoff vector (and coalition structure) for such a situation. Denote by $\widehat{m}(\pi, i)$ this payoff vector, where π is the exiting substructure and i is the arbitrarily chosen first proposer; it is defined on the active player set $S(\pi)$.

I now introduce a condition that combines [M*] and [NAW]. We will need to define affected and unaffected players just as we did for [M*]. To this end, consider a substructure π, and apply the limit algorithm from this point onwards. Fix a player $j \in S(\pi)$. Let T_j denote the maximizing coalition (unique, by [U*]) that contains j. Look at $\pi' \equiv \pi.T_j$, suppose that it is also a substructure, and study the players in $S(\pi')$. Say that player $i \in S(\pi')$ is *unaffected by j* if the maximizing coalition that contains i is exactly the same at the substructures π and π', if the same is true of all other members k of i's maximizing coalition, true of all other members of the maximizing coalitions for all such k, . . . *and* if the worths of *all* these maximizing coalitions are completely unchanged across π and π'.[16] Player i is *affected by j* if she is not unaffected by j.

I can now state the condition:

[N/M] For every substructure π and every pair of active agents i and j in $S(\pi)$ such that i is affected by j, $\widehat{m}_i(\pi, i) > \widehat{m}_i(\pi, j)$.

Just like Condition M*, this is a condition that can be checked directly at the limit, and needs no discount factor. Indeed (and for the same technical reasons as in [M*]), we ask for a strict inequality in the condition. At the same time — and just as in [M*] once again — it would be too much to ask a strict change for *every* player, for some of them may simply be unaffected by the change in initial proposer. This is why the unaffected players need to be excluded.

To interpret this condition, restrict attention first to symmetric games. Consider the algorithm for this class of games, introduced in Section 5.2 of Chapter 5. It involves average worth maximization at every stage, so that $\widehat{m}_i(\pi, i)$ coincides precisely with this maximized average value, which we denoted by $a(\pi)$. What about $\widehat{m}_i(\pi, j)$? This may be viewed as the algorithmic payoff to i when someone else, j, gets to form her coalition first. In the symmetric case, this is just average worth at a further step down the algorithmic line. So, for symmetric games, our condition reduces to a stronger form of NAW: one in which average worth *strictly* declines.[17]

Now consider asymmetric games without externalities. Then it is immediate that [M*] implies [N/M]. In fact, the former condition is actually stronger: it is imposed on *all* pairs of coalitions S and S' with the latter a subset of the former. In contrast, [N/M] only applies to pairs S and S' such that the latter is "reachable" from the

---

[16]This definition can be given a recursive formulation just as in the development of [M*], but it isn't necessary to do so.

[17]As the careful reader of Chapter 5 will have noted, we were nevertheless able to make substantial progress with just the weak inequality. In turn, this allowed us to consider applications that were natural and simple in the symmetric case. However, hidden behind the simplicity is a significant increase in the complexity of the proofs. Given the already complicated arguments for asymmetric games, I did not consider the extra exercise worth the resulting gain in generality for the broader class of games.

former via the elimination of some maximizing coalitions given by the algorithm.

More generally, condition N/M declares that there is no gain in algorithmic value to any player if that player is removed from her role as initial proposer and another agent substituted in her place. This is the assumption that guarantees that a no-delay equilibrium must exist:

PROPOSITION 7.14. *Impose the genericity restriction [U*], and suppose that [N/M] holds. Then there exists a threshold $\delta^* \in (0, 1)$ such that for all $\delta > \delta^*$, a unique no-delay equilibrium exists; it is described in Proposition 7.13.*

If you've accompanied me through the arguments up to this point, you will find the proposition intuitive: intuitive enough to roll a proof and discussion into one. Suppose that everyone follows a no-delay strategy. Then we know what the next formed coalition will be after any substructure: simply find the next proposer according to the protocol, and (provided that the discount factor is close to 1) invoke the unique coalitional maximizer from the limit algorithm that contains this proposer. This means that a particular proposer knows precisely how every substructure that he generates (by means of a proposal) will be "completed" by the remaining players. This creates for our proposer at substructure π the very same characteristic function that our limit algorithm constructed, which is $v_{\Lambda\pi}$.

Construct the **m**-vector corresponding to this characteristic function. By the assumption that everyone follows the no-delay strategy, all respondents must use their component of the vector as their response threshold. Therefore our proposer is left with only two choices: to make an acceptable offer, or to make an unacceptable offer as in Example 7.2 and settle into a subgame where (by assumption) the no-delay strategy will continue to be followed. By undertaking the latter course of action, our proposer — call her i — will receive (approximately, for δ close to 1) a value of $m_i(\pi, j)$, where j is the one who rejects i's unacceptable offer. By sticking to acceptable proposals, on the other hand, our proposer will earn (approximately) $m_i(\pi, i)$. Now, both these payoffs are approximations because they pertain to the limit as $\delta \to 1$, but Condition N/M guarantees that the latter payoff must dominate the former for δ high enough. We have therefore proved that a proposer will indeed follow *her* component of the no-delay equilibrium when everyone else is doing the same. This settled, it is easy to check that a responder will abide by her prescription as well.

Therefore no one-shot deviation from the no-delay strategy profile is profitable, and we are done.

7.7 Alternative Protocols

The analysis in this chapter relies on two assumptions made about the bargaining protocol. First, an initial proposer is deterministically attached to each active player set. Second, the first rejector of a going proposal gets to make a new proposal. How sensitive are the results to the protocol? And how applicable is the methodology developed in this chapter to alternative protocols?

Begin with the first assumption, that an initial proposer is deterministically assigned to each active player set. In symmetric games, studied in Chapter 5, the identity of such a proposer is generally irrelevant, and in particular it does not matter whether or not that proposer is chosen in some deterministic way. The same is true for characteristic function games, but for different reasons: the worth of a formed coalition is independent of other coalitions, so once again it does not matter who proposes after the coalition is formed, or indeed whether this proposer is chosen deterministically or otherwise.

But the identity and ordering of initial proposers do matter in the general case, in which there are asymmetries across players and externalities across coalitions. And indeed, they *should* matter. Player 1 may want to form coalition A, and player 2 coalition B, so whether A or B form may depend on who gets to make a proposal first. Indeed, a theory that purports to yield solutions that are independent of proposer ordering is suspect. The main point is that our methodology works perfectly for any deterministic proposer order, though the predicted equilibrium structure will change (as it well should).[18]

---

[18]If initial proposers are stochastically chosen, the analysis is a bit messier but the same methods work. For instance, consistent coalition maps will now have to be stochastic and the formation of the "artificial" characteristic function at each step of the algorithm will now require us to take expected values over various possible realizations of coalitional worth (depending on precisely how a substructure will be finally completed). This is certainly more complicated, at least in an expositional sense. But the same method works: an **m**-vector is still defined for that characteristic function, no matter how it is constructed. And an

Our second assumption concerns the rejector-proposes protocol. Intuitively, it would appear that when a rejector is given less power following her rejection, she can be "exploited" more easily by the proposer. In turn, this means that a larger share of the social surplus from the formation of a coalition accrues as *private* surplus to the proposer, so that she is more likely to make an efficient offer.

This intuition is not incorrect, though it can be misleading. Recall that for symmetric games, the uniquely predicted coalition structure from the rejector-proposes protocol was immune to a wide range of alternative protocols. The reason for this is that one could — for symmetric games — always construct a class of equilibria in which the rejector is included in every counter-proposal, whether or not she makes that proposal. When the rejector is so included, the result is to imbue her with a lot of effective power, equivalent to that under the rejector-proposes protocol.[19] On the other hand, there are equilibria in which a rejector might systematically be *excluded* whenever she does not get to counter-propose, and these equilibria may enhance efficiency (though at the cost of within-coalition equity).

Example 5.2 in Chapter 5 makes some of these points for symmetric games with externalities, but we can develop the idea in a more illuminating way for characteristic functions. Consider the following example:

EXAMPLE 7.8. *A Symmetric Game With Alternative Protocols.* $N = \{123\}$, $v(i) = 0$ *for all* i, $v(\{ij\}) = a$ *for all* $\{ij\}$, *and* $v(N) = b$, *where we assume that* $b > a > 0$. *This game is strictly superadditive.*

Expand the protocol as follows: a rejector gets to propose with probability p, while with probability $1 - p$ some other active player is chosen equiprobably. Suppose initially that each proposer only makes an offer to a two-person coalition (we will presently verify the condition under which this is true). Denote by m the expected payoff to a rejector and by x the payoff to a proposer. Then, if a rejector is always excluded from a counterproposal by her compatriots, we

[19]To see this why this happens, recall our description of Rubinstein–Ståhl bargaining in Section 4.4 of Chapter 4. As bargaining frictions vanish, equilibrium payoffs in that model converge to equal division no matter how small the odds that the rejector gets to counterpropose. Note that the rejector is included in every counterproposal, because only one coalition — the grand coalition — has positive worth in the Rubinstein–Ståhl model.

have
$$m = \delta px, \text{ while } x = a - m.$$
Solve this to see that
$$m = \frac{\delta pa}{1 + \delta p},$$
which converges to $ap/(1 + p)$ as $\delta \rightarrow 1$. Now, the presumption of a two-person equilibrium coalition is justified if the inclusion of a third person brings a lower surplus gain $b - a$ than the amount needed to pay her off $(ap/(1 + p))$. So for there to be an (inefficient) equilibrium under low enough bargaining frictions in which the last rejector is invariably excluded, the condition

$$b < a\frac{1 + 2p}{1 + p}$$

is necessary and sufficient. When $p = 1$, we are back to the rejector-proposes protocol and the condition reduces to $b < (3/2)a$, which lines up precisely with our characterization of strong efficiency. In the other direction, we have the uniform protocol, in which a rejector has the same chances as anybody else to become the next proposer. Now the condition reduces to $b < (5/4)a$. A *lower* degree of superadditivity is needed to force ubiquitous inefficiency.[20]

Now let's play the flip side: what happens when this condition is *not* met? Do we then get full efficiency in every equilibrium? Suppose that we do. Then every proposer must make an acceptable proposal to the grand coalition, and by a previous argument, the payoffs of every player must converge to equal division — $b/3$ — as bargaining frictions vanish. But now, provided that $b < (3/2)a$ (which allows, of course, for b to exceed $(5/4)a$), a proposer can make an acceptable offer to a two-person coalition which yields strictly more than b/n to each member, herself included. This is a contradiction. Therefore *full* efficiency is not possible.

The argument in the previous paragraph is perfectly general, as Okada (1996) has shown. Provided that there is some coalition S with average worth that exceeds that of the grand coalition, a proposer in S can engineer a profitable deviation from a putative equilibrium in which everyone makes an offer to the grand coalition

---

[20]Don't forget: there is still an equilibrium (in which the rejector is included) which looks just the same as the one under the rejector-proposes protocol. That's a consequence of our earlier results for symmetric games, and I won't dwell on these any further.

alone. In a sense, then, we reclaim Proposition 7.7 no matter what the protocol is.

Summing up: we haven't strayed too far from the analysis conducted for the rejector-proposes protocol. Under the same conditions as before, equilibria still fail to be strongly efficient, even if new proposers fully exclude the rejector (thus dragging down her outside option). Indeed, if the extent of superadditivity is low enough (but still positive), *every* proposer makes an inefficient offer in every equilibrium, whether or not the rejector is included in the counteroffer.

As a transition to my final point in this section, consider this question for Example 7.8: when $(5/4)a < b < (3/2)a$, the equilibrium cannot be "fully" efficient by Okada's theorem. But how "close" to efficiency can we get? It should be intuitive that the "most efficient" equilibrium must involve the exclusion of the rejector from a counterproposal whenever possible. To this end, suppose that an offer is made to the grand coalition by every proposer with (interior) probability θ, and only in this case is the rejector included by another proposer. Then

$$m = \delta[px + (1 - p)\theta m],$$

but it must also be that m precisely equals $b - a$, otherwise no proposer would be indifferent between making an offer to a two-person coalition and to the grand coalition. We must therefore also have $x = a - m$ (regardless of whether the proposal is made to a to- or a three-person coalition). It follows that as $\delta \to 1$, the limit probability θ satisfies

$$b - a = p(2a - b) + (1 - p)\theta(b - a),$$

or

$$\theta = \frac{b - a - p(2a - b)}{(1 - p)(b - a)}.$$

When $p = 1/3$ so that proposer probabilities are always uniform regardless of who rejects,

$$\theta = \frac{4b - 5a}{2(b - a)},$$

and now we have an answer to our little mystery: when $(5/4)a < b < (3/2)a$, it is possible to get efficiency with probability $(4b-5a)/2(b-a)$. This probability converges to 0 as $b \to (5/4)a$, and to 1 as $b \to (3/2)a$.

For some asymmetric games such mixed-strategy equilibria may acquire even greater cogency, simply because no other equilibrium might exist. To see this, recall the employer–employee game studied in Example 7.6. Remember in that game that $N = \{123\}$, $v(i) = 0$ for all i, $v(\{23\}) = 0$, $v(\{12\}) = v(\{13\}) = 1$, and $v(N) = 1 + \mu$ for some $\mu \in (, 1/2)$. The interpretation is that player 1 is an employer who can produce an output of 1 with any one of the two employees 2 and 3. He can also hire both employees in which case output is higher. No other combination can produce anything.

Alter the protocol by supposing that the first rejector of a proposal gets to propose with probability p, while the remaining probability is divided equally among the remaining two players. As before, begin by assuming provisionally that all parties make proposals to two-person coalitions. Notice that player 1 is included in every coalition. However, let us make sure to exclude players 2 and 3 if they reject a proposal and they are not called upon to propose next. Then both players 2 and 3 have a threshold of

(7.15) $$m \equiv m_2 = m_3 = \delta p(1 - m_1),$$

while

$$m_1 = \delta\left[p(1 - m) + (1 - p)m_1\right].$$

Solving for m_1 and substituting this result in (7.15), we see that

$$m = \frac{\delta p(1 - \delta)}{1 - \delta(1 - p) - \delta^2 p^2},$$

so that m converges to zero as bargaining frictions vanish, no matter how close p is to 1. This is quite different from the rejector-proposes scenario, in which p *equals* 1, and m converges to $1/2$ as $\delta \to 1$.

But now we must recall the possibility that three-person coalitions can also form. If so, player 2's (or 3's) response threshold *cannot* go to zero, for if it did everybody would be making acceptable proposals to the grand coalition. But then we start going around in circles, for if everybody made acceptable proposals to the grand coalition each of their m-values would converge to $(1 + \mu)/3$, but because we assumed that $\mu < 1/2$ this can't happen either (agents would then only make two-person offers). This informal discussion suggests that a pure-strategy equilibrium does not exist, and that the players must want to propose to the grand coalition with some probability between zero and 1. That, in turn, must mean that m equals μ, and pins down the required probability θ that a proposal to the grand coalition will be made by a rejector's compatriots, For players 2 and 3, we must have

$$m = \mu = \delta \left[p(1 - m_1) + \theta(1 - p)\mu \right].$$

For low bargaining frictions, it can be seen that $\theta < 1$, so we do not run afoul of Okada's proposition that strong efficiency must fail in this example. Yet, as $\delta \to 1$, θ converges to 1 and the equilibrium can yield efficient outcomes with very high probability.

It is clear from this discussion that the study of alternative bargaining protocols is bound to lead to new and interesting insights. Some of these insights are fully in line with the rejector-proposes protocol, though the quantitative implications may differ. Others may be more radically different, as our study of the employer–employee game suggests. Yet in all cases it appears that the general methodology introduced in this chapter is useful. Whether or not this assertion receives firmer support must be the support of future research.[21]

7.8 Proofs

Proof of Observation 7.3. For i and T as described in the statement of the lemma, we have that

$$(7.16) \qquad y_i(S, \delta) = \delta \left[v(T) - \sum_{k \in T - i} y_k(S, \delta) \right],$$

while for $j \in T - i$,

$$(7.17) \qquad y_j(S, \delta) \geq \delta \left[v(T) - \sum_{k \in T - j} y_k(S, \delta) \right].$$

Adding $-\delta y_j(S, \delta)$ to both sides of (7.17) and using (7.16), we see that

$$(1 - \delta) y_j(S, \delta) \geq \delta \left[v(T) - \sum_{k \in T - j} y_k(S, \delta) \right] - \delta y_j(S, \delta) = (1 - \delta) y_i(S, \delta).$$

$\square$

---

[21]As this book was going to press, Akira Okada sent me a draft of his new paper (Okada (2007)) which studies mixed strategy equilibria under the random-proposer protocol. He also brought to my attention the paper by Kawamori (2006), which studies the mixed protocol discussed in this section.

Proof of Proposition 7.2. First we show that the description in the proposition constitutes an equilibrium. Inductively, assume that it is true for all active player sets of cardinality $s - 1$ or less (indeed, it is trivially true when $s = 2$). Now pick a set S (with cardinality s) of active players, and any proposer $i \in S$. By following the equilibrium prescription, our player can earn a payoff of $m_i(S, \delta)/\delta$.

Suppose that he deviates by making an alternative proposal; then that new proposal must be unacceptable.[22] By applying the going strategy profile from this point onwards, i's subsequent present-value payoff is bounded above by $\max_{T \subseteq S} m_i(T, \delta)$.[23] By Condition M, such a deviation cannot be profitable.

It is very easy to check that prescribed responder strategies are best responses as well.

To show that nothing else can be an equilibrium, proceed inductively once again. Suppose that uniqueness holds for all active player sets of cardinality $s-1$ or less (this is trivially true when $s = 2$). Now pick a set S of cardinality s. Consider any equilibrium response vector $\mathbf{y}$ on S. Let

$$K \equiv \{i \in S | m_i(S, \delta) \neq y_i\}$$

and pick an index i such that either m_i or y_i is the *biggest* of the values featured in K. If it is y_i, note that $y_i > m_i(S, \delta) \geq \delta m_i(S', \delta)$ for all $S' \subseteq S$ (by Condition M). Notice that player i's payoff as proposer is y_i/δ, so the previous inequality and the induction hypothesis jointly imply that i *must be making an acceptable offer.* So pick T such that

$$(7.18) \qquad y_i(S, \delta) = \delta \left[v(T) - \sum_{k \in T-i} y_k(S, \delta) \right],$$

and observe that

$$(7.19) \qquad m_i(S, \delta) \geq \delta \left[v(T) - \sum_{k \in T-i} m_k(S, \delta) \right].$$

---

[22]Given that everyone else is using the response vector $\mathbf{m}$, the prescribed payoff is the most that i can get by making an acceptable proposal.

[23]If, following the unacceptable offer, i is included in the very next proposal, he will get a present value of no more than $m_i(S, \delta)$. If not, he will at best be proposer in some subsequent subset T of active players, whereupon by induction he receives $m_i(T, \delta)/\delta$. However, this payoff is discounted by *at least* δ because of the delay caused by the unacceptable proposal.

By (7.18) and Lemma 7.3, $y_j \geq y_i$ for all $j \in T$. So $m_j(S, \delta) \leq y_j$ for all $j \in T$, $j \neq i$. But then using (7.19), we see that $m_i(S, \delta) \geq y_i$, which is a contradiction.

Alternatively, if at the index i in the "maximal set" K we have $m_i(S, \delta) > y_i$, then note that there exists T such that (7.18) holds with equality with m_i in place of y_i, and then follow the same argument (from (7.18) onwards) with the roles of m_i and y_i interchanged. $\square$

Proof of Proposition 7.3. Pick $i \in S$, and suppose that i is assigned a value m_i at Step $K + 1$ of the algorithm. Suppose that W (of the form $C_{K+1} \cup T_{K+1}$) is the algorithmic maximizer at that step, where C_{K+1} and T_{K+1} are as described in the algorithm. Then

$$m_i = \delta \left[\frac{v(W) - \sum_{j \in T} m_j}{1 + \delta(c - 1)} \right]$$

$$= \delta \left[v(W) - \sum_{j \in W - i} m_j \right],$$

where the equality follows from simple transposition of terms.

It remains to show that for every other coalition W with $i \in W$

$$m_i \geq \delta \left[v(W) - \sum_{j \in W - i} m_j \right].$$

Suppose this is false. Then there is W such that $i \in W$ and

(7.20)
$$m_i < \delta \left[v(W) - \sum_{j \in W - i} m_j \right].$$

Pick $k \in W$ such that k belongs to the "highest" index in the algorithm, say ℓ. Then by construction, $m_k \leq m_i$, so that in combination with (7.20),

$$m_i + (1 - \delta)(m_k - m_i) < \delta \left[v(W) - \sum_{j \in W - i} m_j \right],$$

and rearranging, this tells us that

(7.21)
$$m_k < \delta \left[v(W) - \sum_{j \in W - k} m_j \right].$$

Now, as far as k is concerned, W is a set of the form $C \cup T$ at Step ℓ, and it is also true (by choice of k) that $m_j = m_k$ for all $j \in C$. Therefore (7.21) implies that

$$[1 + \delta(c - 1)]m_k < \delta \left[v(W) - \sum_{j \in T} m_j \right],$$

which contradicts algorithmic maximization at Step ℓ.

Proof of Proposition 7.4. We proceed by induction on the steps in the algorithm. First consider Step 1. Look at the values generated by two distinct coalitions C and C', at least one of which is a nonsingleton; they are

$$\frac{v(C)}{1 + \delta(c - 1)} \quad \text{and} \quad \frac{v(C')}{1 + \delta(c' - 1)}.$$

Given [G.1], these values can never be equal if $c = c'$. If $c \neq c'$, equality can hold for at most *one* value of δ. It follows that there exists $\delta_1 \in (0, 1)$ such that if $\delta \in (\delta_1, 1)$, the maximization problem in Step 1 is solved either for a *unique* coalition (if it is nonsingleton) or for one or more singleton coalitions. Moreover, the solutions are uniform over δ in this interval.

Recursively, suppose that for Steps $1, \ldots, K$, there exists a threshold $\delta_K \in (0, 1)$ such that the properties described in the previous paragraph are true for every $k \in \{1, \ldots, K\}$. Consider Step $K + 1$. Look at the values generated by two distinct coalitions $C \cup T$ and $C' \cup T'$, at least one of which is a nonsingleton; they are

$$\frac{v(C \cup T) - \sum_{j \in T} m_j}{1 + \delta(c - 1)} \quad \text{and} \quad \frac{v(C' \cup T') - \sum_{j \in T'} m_j}{1 + \delta(c' - 1)}.$$

Consider the difference; this is

$$\frac{v(C \cup T) - \sum_{j \in T} m_j}{1 + \delta(c - 1)} - \frac{v(C' \cup T') - \sum_{j \in T'} m_j}{1 + \delta(c' - 1)}.$$

By the induction hypothesis, this is a polynomial expression in δ over the domain $(\delta_K, 1)$. Such an expression *either* admits a finite number of values for δ for which it equals zero, *or* it is entirely independent of δ and equals 0 throughout. In the latter case one can artificially extend the expression over all of $[0, 1]$ by keeping each of the coalitions in the previous steps unchanged at their solutions for $\delta > \delta_K$. Setting $\delta = 0$, we should then have

$$v(C \cup T) - v(C' \cup T') = 0,$$

which is impossible given [G.1]. This means that that former case is the one that applies, which proves the inductive step for $K + 1$: a threshold $\delta_{K+1} \geq \delta_K$ can be found with the required property.

Because there are only a finite number of steps, take δ^* to be the largest of all the δ_K's; this lies between 0 and 1. For $\delta > \delta^*$, pick any proposer i. Suppose that i belongs to set C_K of the algorithm. If that step prescribes a nonsingleton set, we are done. If it prescribes (perhaps several) singleton sets, we know that she can choose only one of them — the set $\{i\}$. □

Proof of Proposition 7.6. First we show that [M] must hold for all discount factors sufficiently close to 1. By the assumed uniqueness of maximizing coalitions under the limit algorithm, for each active player set S there exists a threshold $\delta(S) \in (0,1)$ such that the algorithm yields the very same maximizing coalitions as the limit algorithm (for the same active player set) at every discount factor $\delta > \delta(S)$. Define δ^1 to be the largest of these thresholds as we range over all active player sets; then $\delta^1 \in (0,1)$ as well.

Now consider the corresponding values $\mathbf{m}(S, \delta)$. These converge to $\mathbf{m}^*(S)$ for every active player set as $\delta \to 1$. Thus, given [M*], for any pair of active sets S and S' with $S \supseteq S'$, there exists a threshold $\delta(S, S') \in (\delta^*, 1)$ such that for all affected $i \in S'$,

$$m_i(S, \delta) \geq m_i(S' \delta)$$

for all $\delta > \delta(S, S')$, while for all unaffected i,

$$m_i(S, \delta) = m_i(S' \delta)$$

for all $\delta > \delta(S, S')$. Pick δ^* to be the maximum of all such $\delta(S, S')$. Then $\delta^* \in (0,1)$, and [M] holds for all $\delta > \delta^*$.

The remainder of the proof simply consists in applying Proposition 7.2, and noting that for all $\delta > \delta^*$, the algorithm yields a unique maximizing coalition at every stage for every active player set. □

LEMMA 7.1. *Fix an equilibrium. For any (S, δ), suppose that (7.2) holds with strict inequality for some i:*

$$y_i(S, \delta) > \delta \max_{T : i \in T \subseteq S} \left[v(T) - \sum_{j \in T - i} y_j(S, \delta) \right].$$

Then there exists a strict subset S' of S such that $y_i(S', \delta) \geq y_i(S, \delta)$.

Proof. Let S^* be some *minimal* subset of S (which could be S itself) such that $y_i(S^*, \delta) \geq y_i(S, \delta)$. Then I claim that

$$y_i(S^*, \delta) = \delta \max_{T:i\in T\subseteq S^*} \left[v(T) - \sum_{j\in T-i} y_j(S^*, \delta) \right].$$

For if not, then i must make an unacceptable proposal at S^*. But after that he can get *at most* $y_i(S', \delta)$ for some $S' \subset S^*$. But by construction, $y_i(S', \delta) < y_i(S^*, \delta)$, which is a contradiction. So equality does hold, which means from our premise that S^* must be a strict subset of S.   □

Proof of Proposition 7.7. First we show that (7.9) implies strong efficiency. Pick any person i and look at $y_i(N, \delta)$. Then, using Lemma 7.1, there exists S (could be N itself) and $T \subseteq S$ with $i \in T$ such that

$$(7.22) \qquad y_i(N, \delta) \leq y_i(S, \delta) = \delta \left[v(T) - \sum_{j\in T-i} y_j(S, \delta) \right].$$

We know from Observation 7.3 that $y_j(S, \delta) \geq y_i(S, \delta)$ for all $j \in T$, so using (7.9),

$$(7.23) \qquad y_i(S, \delta) \leq \frac{\delta v(T)}{1 + \delta(t - 1)} \leq \frac{\delta v(N)}{1 + \delta(n - 1)}$$

where t and n are the cardinalities of T and N respectively.

Note that *strict* inequality must hold in the second inequality of (7.23) whenever $T \neq N$.

So, combining (7.22) and (7.23), we have shown that

$$(7.24) \qquad y_i(N, \delta) \leq \frac{\delta v(N)}{1 + \delta(n - 1)}$$

for all i, with strict inequality whenever i does not propose acceptably to the grand coalition.

At the same time, we know that

$$(7.25) \qquad y_i(N, \delta) \geq \delta \left[v(N) - \sum_{j\in N-i} y_j(N, \delta) \right]$$

$$(7.26) \qquad \geq \delta \left[v(N) - \frac{\delta v(N)(n - 1)}{1 + \delta(n - 1)} \right]$$

$$(7.27) \qquad = \frac{\delta v(N)}{1 + \delta(n - 1)},$$

where the second line uses (7.24). Together, (7.24) and (7.27) prove that i must make an acceptable offer to the grand coalition, which proves efficiency.

Conversely, suppose that we have strong efficiency for all discount factors close to 1. Then $y_i(N, \delta)$ is some constant y for all i, *and*

$$y_i(N, \delta) = \delta \left[v(N) - \sum_{j \in N-i} y_j(N, \delta) \right],$$

so that

(7.28)
$$y = \frac{\delta v(N)}{1 + \delta(n - 1)}.$$

Moreover,

$$y_i(N, \delta) \geq \delta \left[v(S) - \sum_{j \in S-i} y_j(N, \delta) \right]$$

for all S, so that

(7.29)
$$y \geq \frac{\delta v(S)}{1 + \delta(s - 1)}.$$

Combine (7.28) and (7.29), and send δ to 1. □

Proof of Proposition 7.13. The argument pieces together several propositions in the text. Begin by making the inductive assumption that the proposition is true for all partition functions with player sets of cardinality $n - 1$ or less. This presumption is trivially true when $n = 1$. Now pick a partition function with player set N (of cardinality n). Our induction hypothesis immediately guarantees that the proposition is true at every stage for which some coalition has exited and the game is about to continue for the remaining set of active players. This is because a partition function is induced for all such remaining sets in the obvious way. Because there are only finitely many such partition functions, there is a single threshold — call it $\hat{\delta}$ — such that the predictions of the proposition hold for all these partition functions.

Fix any $\delta > \hat{\delta}$. Pick any initial proposer $i \in N$. By induction, if i forms the coalition S, the resulting full coalition structure will surely form by applying the consistent completion map $c(S, \Lambda)$ to the substructure with single coalition S. Therefore the worth of coalition S is unambiguously defined by

$$v_{\Lambda\phi}(S) = v(S, c(S, \Lambda)).$$

This is a well-defined characteristic function. Therefore, no-delay equilibrium response thresholds must be characterized by

$$(7.30) \qquad y_i(N, \delta) = \delta \max_{S:i \in S \subseteq N} \left[v_{\Lambda\phi}(S) - \sum_{j \in S-i} y_j(N, \delta) \right]$$

for every i, which just means that the thresholds must be given by the **m**-vector corresponding to the characteristic function $v_{\Lambda\phi}$ for the player set N with discount factor δ. These values must converge in turn to the limit **m***-vector for the associated characteristic function $v_{\Lambda\pi}$. Moreover, by [U*], the algorithm used to compute the limit **m***-vector admits a unique coalitional solution at any step. It follows that there exists $\delta^* \geq \hat{\delta}$ such that for all $\delta > \delta^*$, player i's choice of coalition is given precisely by the unique coalition containing i in Step 4 of the limit algorithm for the general case.

Combining these observations, we are done with the inductive step, and the proof of the proposition is complete. □

7.9 Summary

In this chapter, which is necessarily long and fairly complex in its exposition, we study coalition formation for heterogeneous partition function games. The symmetric case studied earlier is a useful guide but several new considerations arise. For this reason we proceed in two main steps.

The first step consists in developing the theory for general characteristic functions; that is, for situations in which there are no externalities. We allow for arbitrary degrees of heterogeneity across players. The emphasis is on developing a workable theory for no-delay Markovian equilibria, in which all proposers make acceptable offers at every stage of the game. It turns out that the equilibrium payoffs for such games have a remarkably clean and simple structure: they are unique for each set of active players, and moreover they are amenable to computation by the use of a recursive algorithm. We spend a fair amount of time developing this theory, and discussing the efficiency properties of such equilibria. Most importantly, we lay down a set of conditions (summarized as Condition M) under which every Markovian equilibrium must, indeed, be no-delay.

The second step extends the theory to cover externalities across coalitions. The idea of a *completion map* is central to this extension. Loosely speaking, a completion map converts every partial coalition structure into a full coalition structure for the game. It embodies an equilibrium continuation that must, in the sequel, be solved for in its own right. Setting aside this issue for the moment, the existence of some — any — completion map implies that a characteristic function can be suitably defined on the remaining set of active players once some substructure of players has departed. On this characteristic function we bring to bear the theory summarized in the preceding paragraph. The circle is fully closed by requiring that the "next" coalition predicted by this theory coincides with the "next" coalition predicted by the completion map.

We therefore obtain a theory of coalition formation when commitments to form coalitions are irreversible. Our next task is to study the case in which commitments can be reversed, although only with the consent of all contracting parties.

CHAPTER 8

A Framework for Reversible Agreements

So far, we have assumed that a commitment to form a coalition, once made, cannot be undone. As we have argued, in many situations this isn't a bad assumption. Often commitments cannot be made on legal paper; they are executed through the taking of physical actions that are costly to reverse. For instance, in the pollution control example studied in Chapter 6, the costs of reversing a "commitment to not participate in pollution control" may be low if anti-pollution devices can be tagged on to factories in a modular way. If, on the other hand, such factories must be rebuilt from scratch, the cost of reversing the no-control commitment can be extremely high. Or consider the decision of a group to start a conflict, or a region to secede: once under way, these decisions are hard to reverse.

But my intention is not to defend the irreversible commitments framework without qualification. There is a host of situations where agreements are binding, but may be reversed freely (or at low cost). A free-trade area or customs union may initially exclude certain countries and later incorporate them. While two firms might merge, a multi-product firm may also spin off divisions into sub-firms. Political coalitions may form and reform.

The purpose of this chapter is to study a model of binding agreements, in which fresh proposals can always be made, and existing arrangements dissolved (with the consent of the signatories to those agreements). Thus coalition formation occurs in "real time", with payoffs received concurrently. I accommodate a wide variety of payoff structures, superadditive or otherwise, transferable or otherwise, and bargaining protocols — an even wider class than considered in earlier chapters. For the main result, I also drop the restriction to Markovian equilibria; I explain the significance of this relaxation below. The analysis draws heavily on Hyndman and Ray (2007).

We continue to keep in mind the benchmark of the Coase theorem, which asserts that in the absence of incomplete information, the outcome of group negotiations should always be efficient. We'll see that the ability to make reversible commitments may take us closer to the efficiency result asserted by Coase, at least "in the long run". As we have already noted, this is far from true in the irreversible commitments case.

8.1 An Example

What is the role played by a reversible commitment? Why would a commitment to form a group first be made, then reversed? Why not simply eschew the making of that commitment in the first place? To see this, I recall the three-player public goods example from Chapter 2, subsequently explored in greater detail in generality in Chapter 6. I permit commitments to be reversed. The purpose of this example is to not only to show why reversible commitments are more conducive to efficiency, but also to argue that the emergence of such efficient outcomes may not be immediate.

EXAMPLE 8.1. *Three symmetric players provide a pure public good. Each unit of contribution r yields one unit of the good, but generates a convex utility cost* $(1/3)r^3$. *Payoffs are transferable across agents, by the use, say, of linearly-valued money.*

In Chapter 2 (and more generally, in Chapter 6) we construct the partition function for this situation:

$$\mathbf{v}(\{123\}) = \{6\sqrt{3}\}$$
$$\mathbf{v}(\{1\}, \{2\}, \{3\}) = \left\{2\frac{2}{3}, 2\frac{2}{3}, 2\frac{2}{3}\right\}$$
$$\mathbf{v}(\{i\}, \{jk\}) = \left\{2\sqrt{2} + \frac{2}{3}, 2\left[1 + \frac{2}{3}\sqrt{8}\right]\right\}.$$

As in the previous analysis, the exact numbers are unimportant. The two critical features are: (a) the per-capita worth of the grand coalition is smaller than the payoff to i in the coalition structure {i}, {jk}; and (b) the per-capita payoff to j and k in the coalition structure {i}, {jk}, exceeds their corresponding payoff in the coalition structure of singletons.

In line with the previous chapters, assume that an initial proposer is drawn randomly, that proposals must be universally acceptable to the players involved, and that the first rejector of a going proposal gets to make a new proposal. Then, if group formation is irreversible, there is only one (numerical) equilibrium structure. Player i stands alone, and players j and k band together. The outcome is inefficient.

Now suppose that proposals can be made on an ongoing basis, even when a full set of coalitions have had the chance to form. Then there are two possibilities, both leading to an efficient outcome. First, if a player moves off on her own, the other two players disband as well, incurring a temporary loss of payoff but thereby getting into position to enforce a symmetric, efficient outcome with the grand coalition forming. If this path indeed constitutes credible play, then no player will move off in the first place, and the outcome is efficient to begin.

Observe that this isn't even a possibility if commitments are irreversible. Once player i moves off, there is no bringing her back, so players j and k will never disband. Observe, moreover, that in this outcome the tying and untying of commitment need not occur "on the equilibrium path", but its very possibility changes the play.

The second possibility concerns a situation in which once player i moves off, players j and k do not find it worthwhile to disband. For instance, this could happen if player i can make a commitment which is irreversible for some length of time, a situation which can be readily modeled by lowering the discount factor of all players. In this case the outcome will still be efficient, but the path to efficiency as well as the final outcome will look very different. Some player i *must* initially move off. Thereafter, players j and k must cajole her back to the grand coalition with an offer that gives her more than what she gets in the structure ({i}, {jk}). By examining the partition function in the example, it is easy to see that an allocation exists for the grand coalition that makes both i as well as the pair jk better off.

So we are ultimately at an efficient outcome, but one that is "skewed" in favor of the individual who was lucky enough to be the first to make a commitment. Notice that *the commitment must have been made* for her to take advantage of it, and so the equilibrium path involves a transitory phase of inefficiency, followed by a Pareto-superior outcome.

It is important to note, in this case, that if unequal division *cannot* be tolerated within the grand coalition, then the outcome could be "inefficient" *even if commitments are reversible*. But I place the word "inefficient" in quotes, for if unequal division cannot be tolerated within the grand coalition, then it is unclear that efficiency should be defined using aggregate payoffs in the first place. The analysis to follow takes this criticism on board, and defines efficiency appropriately.

A different issue (with similar implications) arises if the signatories to the agreement to bring player *i* back into the fold cannot commit to honor this agreement in the future. If — in that future — some other player *j* were to unilaterally desert the agreement and take up the same stance as player *i* did, then there may be little in the situation to induce player *i* to take up the conciliatory offer in the first place. It follows, then, that a general theory of ongoing negotiations may want to contend with *both* agreements that are binding (on all signatories) and agreements that are temporary, in the sense that subsets of the signatories can renege on the agreement in the future. I could study take up the general case here. However, recognizing that infinite book length is not conducive to readership, I will not study temporary agreements in this part of the monograph. To be sure, the framework that I will describe is potentially amenable to the study of such agreements and I return to this possibility in Chapter 13.

It is now time to set up a more formal model.

8.2 A Proposal-Based Model of Coalition Formation

8.2.1 General Specification. A proposal-based model of coalition formation in real time consists of the following objects:

[1] a finite set N of *players*;

[2] a compact set X of *states*, and an infinite set $t = 0, 1, 2 \ldots$ of time periods;

[3] an initial state x_{-1} given at the start of date 0;

[4] a *protocol* describing the choice of proposers — and order of respondents — at each date t, possibly depending on the history of events up to that date;

[5] for each state x and proposed new state y, a collection of subsets $S(x, y)$ that can "approve" the move, with $S(x, x)$ the collection of *all* subsets of N;

[6] for each player i, a continuous one period payoff function u_i defined on X, and a (common) discount factor $\delta \in (0, 1)$.

The basic idea of this model is both general and simple. Following each history, date t begins under the shadow of a "going state" x_{t-1}, in place from the previous date. Using the protocol in [4], a player is chosen to make a proposal. The player proposes a state y, possibly different from the one already in place. The proposal must be made to an "approval committee" — a coalition from $S(x_{t-1}, y)$. (Notice that "no change" needs no approval.) If the proposal is unanimously approved by the approval committee the state moves to y; otherwise it stays at x. This process continues *ad infinitum*. Each player receives payoffs as in [6], where expectations are taken not just over proposer choices but possibly over the stochastic choice of proposal as well. Discounted expected payoffs are added over time to obtain infinite-horizon payoffs: these are well-defined because u_i is obviously bounded on X.

8.2.2 A Variant with Upfront Transfers. The specifications [1]–[6] may be augmented to allow for the possibility of upfront transfers. A proposer proposes a new state (just as in the previous section) *and* a vector of upfront bilateral payments $z = (z_{ij})$, positive or negative in the various components, that sum to zero. These one-time upfront payments are presumably designed to lubricate the implementation of the new state.[1]

To incorporate upfront transfers into the basic setup, we will need to add the following points to items 5 and 6 above:

[5'] The approval committee for the proposed move from x to y with transfers z must consist of all members of the approval committee in

---

[1]Gomes and Jehiel (2005) study this model in a more general context which allows for nonbinding agreements. We discuss their important paper in more detail in Section 10.4 of Chapter 10.

the baseline model, as well as any individual involved in a nonzero transfer.

[6'] Each player's payoffs are quasi-linear. That is, current payoffs for i under the state x and overall transfer received, z, are given by $(1 - \delta)u_i(x) + z$.

Infinite-horizon payoffs are obtained, just as in the baseline model, by adding discounted payoffs over time. Because u_i is continuous and X compact, side-payments can also be taken to come from some compact set for every discount factor, so that infinite-horizon payoffs are well-defined.

We will refer to this as the model (or variant) with *upfront transfers*.

Whether or not the baseline model makes more sense than the variant with upfront transfers is a question that requires a contextual answer. If agents are very patient, the required transfers will need to be extremely large. If agents are liquidity constrained, it would be silly to approximate such situations with a model of unlimited upfront transfers: the no-transfers baseline model will do much better. To be sure, there are other situations, such as negotiations across firms with deep pockets, or across countries, where the upfront transfer scenario may be much more plausible.

In the analysis to follow, we will largely focus on the baseline model. Remember that that model also permits (by suitable interpretation of the state) *ongoing* transfers between members of a coalition. At the same time, the variant with upfront transfers has some particular features that we will take up in Chapter 10.

8.2.3 Some More Structure on States. The specification so far is quite general and can be used in a variety of ways. Our particular emphasis is on permanently binding agreements. I will therefore impose some more structure on the notion of a "state".

Suppose that the underlying one-shot interaction is described by a a partition function, which assigns to each partition π and coalition $S \in \pi$ a set of payoff allocations $U(S, \pi)$. This is more general than what we've done so far, in that payoffs may or may not be transferable under this specification. But we will presume that each element of $U(S, \pi)$ is efficient for S, given the structure π. That is, all potential inefficiencies for S under some given coalition structure can be costlessly done away with by the members of S.[2]

Of course, a characteristic function is a special case in which **U** is independent of π.

Now for the added structure: we suppose that every state x can be expressed as a pair $(\pi, \mathbf{u})$, where π is a partition or coalition structure, and $\mathbf{u}_S \in \mathbf{U}(S, \pi)$ for every coalition $S \in \pi$.[3]

8.2.4 Some Structure on Protocols. Notice that our description of the proposer protocol is extremely general, in that it permits the (possibly stochastic) choice of proposer to depend in arbitrary ways on history. Sometimes we will specialize, of course, studying the now familiar rejector-proposes protocol or the random-proposer variant. But our main result in Chapter 9 will be valid for all protocols that satisfy the following mild technical restriction:

[P] For each i, let H_i be the set of histories after which player i is asked to make a proposal with positive probability. Then this probability is *uniformly* positive on H_i.

All that [P] rules out is the rather arcane possibility that some player may be asked to propose along a sequence of histories with a corresponding sequence of positive probabilities that converges to 0. [P] is satisfied for every reasonable protocol that we can think of, including all deterministic and history-independent random protocols.

8.3 Binding Agreements

In the irreversible scenario studied thus far in the book, the notion of a binding agreement is unambiguous. It is an agreement that can be costlessly implemented once agreed upon. But more is at stake when agreements can be renegotiated or discarded. We will need to specify the groups of individuals that are in a position to *alter* a going agreement. The concept of an approval committee is helpful

---

[2]As we have already noted in Chapter 3, Section 3.2.4, the conversion of a game into partition function form isn't automatic. The very definition presumes a "product structure" in payoffs across coalitions, with each coalition having access to its efficient payoffs (relative to the structure π).

[3]The notation $\mathbf{u}_S$ denotes the projection of the vector $\mathbf{u}$ on S.

for this. Loosely speaking, we would like to say that agreements are *binding* if an individual is on the approval committee for every proposed move that will "affect" an ongoing agreement enjoyed by that individual.

The operative word is "affect", and we break this into two parts. First, if a player's coalitional *membership* is affected as a consequence of a proposed move, the move *must* be disrupting some previous agreement to which that player was a signatory. Existing coalitional membership is, after all, the product of some past agreement. In this case we assume that the individual in question must be on the approval committee for the move.

Second, a proposed move might affect the (ongoing) *payoff* to a particular agent, without altering her coalitional membership. Must consent be sought from that agent? The situation here is more subtle. It may be that the payoff is affected because a fellow-member of a coalition wishes to reallocate the worth of that coalition. In that case — given that the existing allocation is in force — it is only reasonable that our agent be on the approval committee for the move. On the other hand, our agent's payoff may be affected because of a coalitional change elsewhere in the system, which then affects our agent's coalition via an externality. Our agent is "affected", but need not be on the approval committee because she wasn't part of the agreement "elsewhere" in the first place.[4]

We may summarize all this a bit more formally. For any move from x to y, let $C(x, y)$ denote the set of individuals whose coalitional membership is altered by the move, and $P(x, y)$ the set of individuals j whose one-period payoffs are altered by the move: $u_j(x) \neq u_j(y)$. Say that agreements are *binding* if the following restrictions on approval committees are satisfied:

[B.1] For every state x and proposed move y, $C(x, y)$ is a subset of any approval committee for the move.

[B.2] Consider a coalition S with membership entirely untouched by a move from state x to state y. Then, provided either that there has been no change at all in the coalitional structure or that payoffs are described by a characteristic function, every member of $S \cap P(x, y)$ must belong to any approval committee for the move.

---

[4]Notice that we wouldn't insist that our player should *not* be on that approval committee; it's just that our definition of binding agreements is silent on the matter.

The discussion above indicates that [B.2] is the more subtle of the two restrictions. The implicit question that [B.2] asks is simply this: fix a coalition and a move that does not alter membership in this coalition. If, moreover, there is no change in the *entire* coalition structure, or if the situation is describable by a characteristic function to begin with, how could the payoff of a particular agent in the unchanged coalition possibly change? The answer (again implicit in [B.2]) is that it could *only* have changed because there is a deliberate reallocation within that coalition, and then [B.2] demands that all individuals affected by that reallocation must approve the move.[5]

In the variant with upfront transfers, the payoff could also change if there are transfers. But we've already decreed that all individuals receiving or giving transfers must approve the move (see Section 8.2.2, item [5']).

Thus [B.1] and [B.2] formalize binding agreements, and we maintain these restrictions throughout. Sometimes — mainly in the examples — we invoke the *sufficiency* of these restrictions. Say that approval committees are *minimal* if any coalition respecting [B.1] and [B.2] can serve as approval committee for a proposed move.

In closing this section, it should be noted that an entirely different theory may be written down when the restrictions [B.1] and [B.2] are not met. For instance, a theory of "temporary agreements" can be constructed by assuming that agreements only bind for, say, one period. For any move from x to y, any approval committee must contain all members of at least $m - 1$ of the m new coalitions that form, and in particular, must include all new coalitions in y that are not subsets of former coalitions in x.[6]

Chapter 13 will, in fact, study temporary agreements, though it will do so using a blocking framework.

---

[5]Recalling the discussion in Chapter 3, Section 3.2.4, this restriction would make less sense if there were multiple equilibria across coalitions. For then a changed payoff in coalition S could be compatible with no change in the coalition structure if somehow, the selection of across-coalition equilibria were affected by the move. No "deliberate reallocation" within S is involved.

[6]The interpretation is that if a new coalition is formed by taking members from more than one erstwhile group, then all the members of the new coalition must approve the move. At best one coalition may be left out of the approval process, and this coalition must be a subset of an erstwhile coalition. It is to be interpreted as a "residual" left by the other "perpetrating coalitions" (compare with the definition of perpetrators and residuals in Ray and Vohra (1997) and in Chapter 12).

As a second variant, allow a coalition to break up or change if some given fraction (say a majority) of the members in that coalition permit that change. Some political voting games or legislative bargaining would come under this category. Now any approval committee must consist of at least a majority from *every* coalition affected by the move from one state to another.

Going in the opposite direction, [B.1] and [B.2] could be further strengthened: one might require that a coalition once formed can never break up again. This would lead us back to the model with irreversible commitments.

8.4 Strategies and Equilibrium

At each stage of the proceedings, we keep track of past proposers, proposals, rejectors (if any) and upfront transfers (if any). A *history* at some stage of the game is a list of such objects up to, but not including, the events that will occur at that stage. Such stages may be of various kinds: a proposer is about to be chosen, or a proposal about to be made, or a responder about to respond, or — such matters concluded — a state about to be implemented. We use obvious nomenclature to distinguish between the different types: "proposer histories," "responder histories," "implementation histories," and so on.

At proposer or responder histories players have to take deliberate actions. A full listing of a particular player's actions for all such histories is a *strategy* for that player. Notice that we are being deliberately quite general here by allowing for all history-dependent strategies. We'll see why in the next chapter.

To describe strategies more formally, consider an individual k. For a proposer history h at which k is meant to propose, she must choose a (possibly new) state y and an approval committee S for the proposed move. (In the variant with upfront transfers she would also choose a proposed vector of transfers $\mathbf{z}$.) She could employ a behavior strategy, which would be a probability distribution over (y, S). Denote by $\mu_k(h)$ the probability distribution that she uses at proposer history h.[7]

---

[7]Notice that we are allowing any proposer to make a proposal to any committee.

Likewise, at a responder history h at which k is meant to respond, denote by $\lambda_k(h)$ be the probability that k will accept the going proposal under that history. The full collection $\sigma = \{\mu_k, \lambda_k\}$ over all players k is a *strategy profile*.

A strategy profile σ induces *value functions* for each player. These are defined at all histories of the game, but the only ones that we will need to track are those just prior to the implementation of a fresh state (or the unaltered continuation of a previous state). Call these *implementation histories*. On the space of such histories, every strategy profile σ (in conjunction with the given proposer protocol) defines a stochastic process P^σ as follows. Begin with an implementation history. Then a state is indeed "implemented". Subsequently, a new proposer is determined. The proposer proposes a state. The state is then accepted or rejected. (The outcome in each of these last three events may be stochastic.) At this point a new implementation history h' is determined. The entire process is summarized by the transition P^σ on implementation histories.

For each person i and given an implementation history h, the *value* for i at that date is given by

$$(8.1) \qquad V_i^\sigma(h) = (1 - \delta)u_i(x) + \delta \int V_i^\sigma(h')P^\sigma(h, dh')$$

where x is the state implemented at h. Given any transition P^σ, a standard contraction mapping argument ensures that V_i^σ is uniquely defined.

In the variant with upfront transfers the value functions at any implementation history will include the prospect of transfers at future dates, and so will satisfy

$$(8.2) \qquad V_i^\sigma(h) = (1 - \delta)u_i(x) + \delta \int [V_i^\sigma(h') + z_i(h')]P^\sigma(h, dh')$$

where $z_i(h')$ is the overall transfer received by player i following history h'.

In the baseline model, say that a strategy profile σ is an *equilibrium* if two conditions are met for each player i:

(a) At every proposer history h for i, $\mu_i(h)$ has support within the set of proposals that maximize the expected value $V_i^\sigma(h')$ of i, where h' is the subsequent implementation history induced by i's actions and the given responder strategies.

(b) At every responder history for i, $\lambda_i(h)$ equals 1 if $V_i^\sigma(h') > V_i^\sigma(h'')$, equals 0 if the opposite inequality holds, and lies in $[0,1]$ if equality holds, where h' is the implementation history induced by acceptance, and h'' the implementation history induced by rejection.

In the case in which the proposer protocol is history-independent, say that strategies are *Markovian* if h can be replaced by the going state x everywhere in the definitions above. A *Markov equilibrium* is an equilibrium involving Markov strategies.

With upfront transfers the equilibrium conditions must be modified in the obvious way. An individual i as proposer can also announce transfers and seeks to maximize $V_i^\sigma + z_i$ in (a), and as responder uses the net payoff $V_i^\sigma + z_i$ as the appropriate criterion in (b). See Section 10.4 in Chapter 10 for more discussion.

This is a well-defined game of perfect information. Given that X is compact and u_i is continuous for every i, the existence of equilibrium is guaranteed (see, e.g., Harris (1985)). The existence of Markov equilibrium is easy enough to establish if X is finite or countable (see Hyndman and Ray (2007, Supplementary Notes) for details). We omit these relatively technical matters here.

8.5 Absorption and Efficiency

An equilibrium induces a stochastic process on the space of implementation histories. Consider the stochastic process of one-period payoff vectors $\mathbf{u}(x_t)$ thus generated. Say that an equilibrium is *absorbing* if $\mathbf{u}(x_t)$ converges almost surely from every initial state.

A vector of payoffs $\mathbf{u}$ *Pareto-dominates* another vector $\mathbf{u}'$ if $\mathbf{u} \gg \mathbf{u}'$. A payoff vector exhibits (static) *efficiency* if it is not Pareto-dominated by any payoff vector associated with some other state.

We can easily apply this concept to absorbing equilibria with well-defined payoff limits. Specifically, say that absorbing equilibria are *asymptotically efficient* if their payoff limits are (static) efficient.

To be sure, we can be more demanding in our efficiency requirement. Say that an equilibrium is *dynamically efficient* from some initial history h if the vector $\mathbf{V}^\sigma(h)$ is not Pareto-dominated by the infinite-horizon payoff arising from some, conceivably stochastic, sequence of states.

Whether dynamic or static, our notion of efficiency must respect the very same constraints that the players themselves face. In particular, if payoffs cannot be freely transferred across players it would be inappropriate to label an equilibrium as inefficient if it fails to maximize, say, the sum of total surplus. So lack of transferability, for instance, should not be judged as a *prima facie* correlate of inefficiency. The efficiency definition itself must be suitably modified.

8.6 Summary

In this short chapter, we motivate and set up the reversible commitments model. The basic viewpoint adopted here is that a model of coalition formation with ongoing negotiations is best viewed as a dynamic process in which proposals and payoffs are fully intertwined. As new proposals are made and accepted, the going state changes, and payoffs change accordingly. Individuals do not use one-period payoffs to evaluate changes in state; they use the entire continuation value.

Now, while renegotiation is permitted, we impose the restriction that agreements are permanently binding, so that all past signatories to a going agreement must approve any change to it. We discuss a formalization of this idea.

Finally, we introduce notions of absorption and asymptotic efficiency. An equilibrium is absorbing if the one-period payoffs accruing to every player ultimately settles down (instead of cycling or moving around forever). An absorbing equilibrium is asymptotically efficient if that limit payoff is efficient in the sense of static Pareto efficiency. Similarly, one can define dynamic efficiency.

This model has several useful properties. Apart from capturing ongoing renegotiation, it naturally allows for farsighted behavior. The very presumption that individuals use continuation values to judge the wisdom of moving to a new state suggests that they can, and do, anticipate further changes.

In the two chapters that follow, we proceed to a closer examination of this general setup.

CHAPTER 9

Reversible Agreements Without Externalities

In this chapter, we are going to show that the possibility of reversible commitments in games without externalities leads — ultimately — to efficient outcomes. This is in sharp contrast to games with irreversible commitments, in which inefficiency can be endemic even for characteristic functions.

Section 8.1 of the previous chapter tries to make very clear why reversibility may be conducive to efficiency. If commitments are irreversible, the maximal social surplus is not seized because proposer incentives are distorted by the potential loss of control that accompanies a rejected proposal. This means that a proposer must always give some fraction of the surplus away, and a wedge is driven between socially and privately optimal actions.

On the other hand, intuition suggests that if outcomes can be renegotiated, then the already-agreed-upon arrangements safeguard *existing* payoffs against any loss of control from making a fresh proposal. This suggests two things. First, if there is surplus left on the table, then that surplus should eventually be seized and divided in some way among all parties. Second — and somewhat in contrast to the first point — the seizure of that surplus won't generally happen at the very first round. The safeguards may have to be put in place in earlier rounds, necessitating step-by-step progress towards efficiency (and hence a sacrifice of full dynamic efficiency). The example in Chapter 8 makes these points informally. We will develop these ideas more formally in the context of a characteristic function example.

Precursors to the ideas developed in this chapter include Seidmann and Winter (1998), Okada (2000), and Hyndman and Ray (2007) on which much of the discussion is based. The Seidmann– Winter and Okada papers show that asymptotic efficiency may be reinstated if renegotiation is ongoing and coalitions can only expand. These papers — in line with most others on coalition formation — study Markovian equilibria. With history-dependence, additional complications appear. For instance, even simple bargaining games with three or more players are known to exhibit multiple equilibria, many of which are inefficient. Because we want to explicitly discuss and tackle this problem, we've left ample room in the definitions for arbitrary degrees of history-dependence.

9.1 Two Examples

9.1.1 Efficiency, But Not Immediately. Recall Example 7.6 from Chapter 7: the employer-employee game. Three agents come together to form a partnership. Agent 1 is special: she must be included in any partnership with positive value. But agents 2 and 3 — call them the ordinary agents — are needed as well (agent 1 cannot produce value on her own). Write the characteristic function as follows:

$$v(\{1j\}) = 1 \text{ for } j = 2, 3; \quad v(\{123\}) = 1 + \mu,$$

for some $\mu > 0$, and $v(S) = 0$ for all other coalitions.

Here is the protocol: if a proposal has just been rejected, the first rejector proposes. In any other situation, a proposer is chosen at random.

If only irreversible commitments are possible, and the discount factor is close enough to unity, *no agent will ever want to form the grand coalition of all three players, even if the characteristic function is strictly superadditive*, as long as $\mu < 1/2$. The surplus may be highest at the grand coalition, but no one agent has full control over its division. Consequently, a two-person coalition — either {12} or {13} — must form.

What if commitments are reversible? It can be shown that the eventual outcome is indeed efficient (we will prove a general theorem establishing this), but the move to efficiency must be gradual. One of the two-person coalitions {12} or {23} forms first. A subsequent move establishes the grand coalition. But there are unequal returns among the ordinary agents. It can be shown that one of them gets approximately as much as the special agent. But the other agent — the one absorbed later into the group — gets less.

As discussed in the introduction to this chapter, the formation of an intermediate coalition essentially protects the parties to that agreement. The ordinary agent included in the intermediate coalition can block any attempt by the excluded agent to undercut him, because he is already signatory to a binding agreement that can only be abolished with his consent. Because this reduces the power of the excluded agent to extract surplus, the grand coalition can finally form.

We now turn to a second feature that may inhibit efficiency.

9.1.2 History-Dependence Frustrates Efficiency. Perhaps one of the more irritating results of game theory concerns the nihilistic implications of history-dependent strategies. While often used to "explain" the possibility of cooperation in repeated games, history-dependence extracts a price by predicting all sorts of other equilibrium outcomes as well. A well-known expression of all this is the Folk Theorem for repeated games, which shows that all individually rational payoffs may be supported as a perfect equilibrium, provided the discount factor is close enough to one.

When *binding* agreements are studied using the techniques of noncooperative games, the situation is a bit better, but only by a bit. A bargaining game, while possibly involving repeated offers, is *not* a repeated game. (When the bargain is done with, so is the game.[1]) And indeed, the folk theorem does not always apply. For instance, in Section 4.4 of Chapter 4, we studied Rubinstein–Ståhl bargaining and observed that two-person games yield a unique bargaining equilibrium in the class of all strategy profiles, history-dependent or otherwise.

On the other hand, this welcome predictability disappears as soon as there are three or more players. As noted in Chapter 4, the first result along these lines is due to Herrero and Shaked (see Herrero (1985)), in the context of the Rubinstein–Ståhl bargaining model.

---

[1]It is true that we are now allowing for the continued possibility of proposals even after agreements have been written. But our game isn't repeated either: past agreements act as state variables.

Chatterjee, Dutta, Ray and Sengupta (1993) state a folk-theorem-like proposition for all characteristic function bargaining games, not just n-person bargaining games. This literature suggests — along the unsatisfactory lines of the folk theorem — that all sorts of outcomes (including inefficient ones) are possible.

A quick recall of the Herrero–Shaked observation, as well as a variant of it, will be instructive, as it will also serve to motivate an additional assumption to be made below. Suppose, then, that $|N| = 4$, and that the underlying characteristic function is given as follows:

$$v(S) = 3 \text{ if } S = N$$
$$= 0 \text{ otherwise.}$$

The protocol is "rejector proposes".

Provided that the discount factor is close enough to unity, it is possible to sustain an equilibrium outcome in which the grand-coalitional surplus is never attained. If an individual attempts to make a proposal to the grand coalition, that proposal is rejected and every rejector subsequently asks for the entire surplus. Knowing this, no individual makes a proposal to the grand coalition, and the situation stagnates forever in this way.

Of course, matters need not be that dismal. Suppose that we alter this example so that $v(\{12\}) = v(\{34\}) = 1$, and leave all else unaltered. Then there is an equilibrium in which the coalition structure $\{\{12\}, \{34\}\}$ forms but no further progress is made: all additional proposals to the grand coalition are rebuffed in the manner described above (except that the rejector will now ask for the entire surplus net of existing payoffs to the other three agents).

9.2 Benignness

Is it possible to translate the Herrero–Shaked idea to our framework of binding agreements with repeated proposals? Technically, the answer is yes, but the failure to achieve efficiency will be based on rather knife-edge considerations. An efficiency-enhancing proposal may be rejected, true, but events post-rejection cannot hurt our existing players by too much, *because ongoing agreements are binding*. Consider the examples in the previous section. If a proposer does not mind being rejected as long as subsequent play benefits others *and* does not hurt her, such history-dependent inefficiencies can be broken provided that the status quo agreements are binding.

Of course, the example in Section 9.1.2 is only one of many possibilities: history-dependent inefficiencies could, in principle, occur in a variety of ways. Nevertheless, the discussion motivates the following concept: say that an individual is *benign* if she prefers an outcome in which some other individuals are better off, provided that she (and every individual) is just as well off.

The benignness "refinement" is of a lexicographic nature. Our individual first and foremost maximizes her own payoff, and benignness only kicks in when comparisons are made over outcomes in which her payoff is unaffected. There is no danger to the payoff of the individual concerned.

Alternatively, one could just as easily think of benignness as an *equilibrium* refinement rather than as a lexicographic restriction on individual preferences. Thus an equilibrium strategy profile is *benign* if for no individual and no history is there a deviation which increases the payoffs of some players while leaving all other payoffs (including that of the deviating player) unchanged.

9.3 Absorption and Efficiency

This section contains the main results of the chapter. First, all equilibria of characteristic-function games must be absorbing in payoffs. Moreover, provided the benignness refinement is applied, such absorbing states must be efficient, regardless of the degree of history-dependence. We now turn to a precise statement and discussion of these results. The formal analysis is conducted for the baseline model, with some remarks on how to alter the results (if at all) for the variant with upfront transfers.

9.3.1 Absorption. The first important property satisfied by all characteristic functions is that all paths of equilibrium payoffs must "ultimately" settle down.

PROPOSITION 9.1. *Assume* [B.1] *and* [B.2]. *In a game of coalition formation derived from a characteristic function, all equilibria are absorbing.*

It isn't difficult to see why such a proposition must be true. Agreements are binding, as captured by the restrictions [B.1] and

[B.2]. Moreover, characteristic functions don't display externalities. These two observations imply that any change in a player's lifetime value along an equilibrium path must be brought about by her deliberate acquiescence. But this implies the monotonicity of such lifetime values, and hence their ultimate convergence. The associated convergence of one-period payoffs then follows from a simple additional step.

While this proposition is stated (and proved below) for the baseline model, nothing fundamentally different happens in the variant with upfront transfers. Formally, one-period payoffs may not converge simply because there may be ongoing cycles in states supported by transfers back and forth. Such cycles are not of great import: there will always be an equivalent equilibrium path along which the one-period payoffs do converge. I omit these details.

I reiterate that the ability to write binding agreements cannot guarantee *immediate* absorption. As illustrated in Section 9.1.1, equilibrium paths will generally require "time" — i.e., the formation of intermediate coalition structures — before an absorbing outcome (or payoff) is finally arrived at.

9.3.2 Efficiency. Now I turn to the question of efficiency. We have already seen more than one reason to be wary of such an assertion. First, the very jockeying for intermediate handholds of power along an equilibrium path suggests that full *dynamic* efficiency is generally not to be had. We don't have to go very far to verify this: Section 9.1.1 provides an example.

Second, history-dependence and folk-theorem-like arguments might conspire to generate inefficient outcomes even in the long-run. What we are going to show, however, is that benignness rules this possibility out.

PROPOSITION 9.2. *Assume* [B.1] *and* [B.2], *and suppose that the set of states is finite. Then in characteristic function games, every pure strategy benign equilibrium is asymptotically efficient: every limit payoff is static efficient.*

This proposition is particularly remarkable in the light of the folk-theorem-like results obtained in Herrero (1985) and Chatterjee, Dutta, Ray and Sengupta (1993). Under repeated negotiation, we assert that no amount of history-dependence in strategies can hold players away from an (ultimately) efficient outcome. Mainly because we allow for such history-dependence, but also because we include both superadditive and nonsuperadditive cases, this proposition represents a substantial extension of Okada (2000) and Seidmann and Winter (1998), who showed that renegotiation achieves efficiency in superadditive characteristic functions when equilibria are restricted to be Markovian.

A formal proof of Proposition 9.2 is relegated to a later section, but some intuition may be useful. Consider an equilibrium path of play. By Proposition 9.1, the one-shot payoffs along that path converges. Suppose, contrary to our assertion, that convergence occurs to an inefficient limit. Then a proposer will have the incentive to propose a payoff vector that Pareto-dominates this payoff. This follows from two observations. First, because agreements are binding, the proposer cannot be hurt by making such a proposal. She can always continue to enjoy her going payoff. To make this argument work, we must "already" be at the limit payoff, otherwise the proposer may do some (small, but positive) damage to her own prospects by the very act of making the proposal. This is why we work with a finite set of states.

Second, the proposer is benign. She certainly gains from the proposal *if it is accepted*, and there is no reason to invoke benignness. But the point is that she prefers to make the proposal *even if it is rejected*. For rejection must entail that all the rejectors are better off by *not* accepting the proposal, while the assumption that agreements are binding ensures that no one is strictly hurt (see previous paragraph). A benign proposer would therefore prefer the resulting outcome to the presumed equilibrium play, which is continued stagnation at the inefficient payoff vector.

This informal argument is obviously not a complete proof, and several additional points need to be checked. The reader is referred to Section 9.4 for the details.

While we omit a formal analysis, the variant with upfront transfers has no additional insight to offer here. Benignness will guarantee the ultimate attainment of efficiency, with or without upfront transfers.

9.3.3 More on Efficiency. The efficiency proposition does not come for free. We already know that benignness (or something like it) has to be used, otherwise Section 9.1.2 contains a counterexample.

But the statement of the proposition contains other restrictions. We briefly discuss some of the issues here.

9.3.3.1 Static Versus Dynamic Efficiency. It must be reiterated that the ability to write binding agreements cannot guarantee *full* efficiency in the dynamic sense. As illustrated in Section 9.1.1, equilibrium paths will generally require "time" — i.e., the formation of intermediate coalition structures — before an final outcome is finally arrived at. These intermediate outcomes may well be inefficient. So the path taken as a whole cannot be dynamically efficient.

There is another reason for the failure of dynamic efficiency. It is simply that such efficiency may necessitate ongoing cycles across different states. Consider the following example:

EXAMPLE 9.1.

$$x_0 = \{\{1\}, \{2\}\} \quad \mathbf{u}(x_0) = (0, 0)$$
$$x_1 = \{\{12\}, a\} \quad \mathbf{u}(x_1) = (5, 1)$$
$$x_2 = \{\{12\}, b\} \quad \mathbf{u}(x_2) = (2, 2)$$
$$x_3 = \{\{12\}, c\} \quad \mathbf{u}(x_3) = (1, 5).$$

It is easy to see that in any equilibrium, x_1, x_2 and x_3 must all be absorbing states. However, notice that x_2 is an absorbing state which is dominated by randomization between x_1 and x_3, provided that the discount factor is close enough to 1. In a dynamic setting such randomization may be mimicked by alternation between x_1 and x_3. The reason that such inefficiencies cannot be removed by continued negotiation is that the past is not binding on the present *except* via the state, which is restricted to simply reflect the existing coalition structure and payoffs. Once at x_1, say, player 1 will be unwilling to relinquish her highly favorable payoff position.

This example brings up the question: to what extent can history weigh on the present? Can players 1 and 2 write a binding agreement to alternate between states x_1 and x_3? If they can, the inefficiency in the example disappears. One might be tempted to say that such agreements can indeed be written. But similar (though more complex) examples can be constructed in which the domination requires ongoing change in coalition structure and not just the payoff vector. It is unclear where one draws the line.

9.3.3.2 Transferable Utility and Finite State Spaces. Observe that even though our proposition is stated for finite state spaces, we can approximate arbitrarily high degrees of transferable utility. It should therefore not be concluded that our efficiency result is somehow linked to the presence or absence of transferability in payoffs.

The reader may nevertheless wonder if the proposition goes through if the state space is allowed to be infinite. We are not sure of the answer to this question in general, though we would conjecture that it is in the affirmative. For instance, here is a version of Proposition 9.2 when the proposer protocol is restricted to be deterministic.

PROPOSITION 9.3. *Suppose that every individual is benign, the proposer protocol is deterministic and the set of states is compact. Then in characteristic function games with permanently binding agreements, every limit payoff of every pure strategy equilibrium is efficient.*

The proof is omitted; see Hyndman and Ray (2007).

9.3.3.3 Ongoing Negotiations and Benignness. If negotiations are *not* permitted to continue indefinitely, then inefficiency is possible. For n-person bargaining games, where $n \geq 3$, Herrero and Shaked provide the required analysis. The important point is that benignness will do nothing to get rid of such inefficiency.

To see this point clearly, recall the four-person example from Section 9.1.2:

$$v(N) = 3, \quad v(\{12\}) = v(\{34\}) = 1, \quad v(S) = 0 \text{ otherwise.}$$

The protocol is "rejector-proposes". Using arguments similar to Herrero (see Osborne and Rubinstein (1994, p. 130)), and provided that discount factors are close enough to 1, one can easily construct an equilibrium in which players 1 and 2 their worth of 1, while players 3 and 4 divide *their* worth of 1. This outcome is supported by rewarding a player for rejecting a deviant offer by receiving the entire unit pie in the next (and, therefore, every future) period.

This equilibrium is inefficient, and what is more, *the imposition of benignness won't eliminate it.* The reason is that a deviant proposer is *strictly* punished by the above strategies.

In addition, Section 9.1.1 shows that Markovian equilibria may be inefficient as well. These equilibria are also robust to the imposition of benignness. Therefore the assumption of ongoing negotiations is important to the efficiency result.

9.3.3.4 Is Benignness Reasonable? Is the benignness restriction on player preferences reasonable? Obviously, like every assumption it is open to scrutiny. We only mention that benignness has found support in a number of different experimental settings (including bargaining); see, e.g., Andreoni and Miller (2002), Charness and Grosskopf (2001) and Charness and Rabin (2002) among others. Indeed, these studies suggest something stronger: people are sometimes willing to *sacrifice* their own payoff in order to achieve a socially efficient outcome. Given its lexicographic insistence on maximizing one's own payoff, benignness certainly doesn't go that far.

9.4 Proofs

Proof of Proposition 9.1. Fix any equilibrium strategy profile σ and initial condition x_{-1}, and consider the stochastic process on histories thus generated. Conditions [B.1] and [B.2] tell us that for every player i, and for every history h_t with going state x_t,

$$(9.1) \qquad\qquad V_i^\sigma(h_{t+1}) \geq u_i(x_t)$$

for every equilibrium realization of the state h_{t+1} conditional on h_t.

I claim that the induced stochastic process on V_i^σ is a submartingale. To prove this, recall the functional equation

$$(9.2) \qquad V_i^\sigma(h_t) = (1 - \delta)u_i(x_t) + \delta \int V_i^\sigma(h_{t+1})P^\sigma(h_t, dh_{t+1})$$

and use (9.1); it is easy to see that

$$(9.3) \qquad\qquad V_i^\sigma(h_t) \geq u_i(x_t)$$

as well. Now suppose, contrary to our assertion, that

$$\mathbf{E}[V_i^\sigma(h_{t+1})|h_t] < V_i^\sigma(h_t)$$

for some history h_t. Then the functional equation (9.2) implies that

$$V_i^\sigma(h_t) < (1 - \delta)u_i(x_t) + \delta V_i^\sigma(h_t),$$

which directly contradicts (9.3). This proves the claim.

Because V_i^σ is a bounded function on histories, the Martingale Convergence Theorem (see, e.g., Ash (1972, Theorem 7.4.3)) implies that the induced sequence of random variables $V_i^\sigma(h_t)$ converges almost surely to some limit random variable; call it V^*.

Next, observe that the random variable $Z(h_t) \equiv \mathbf{E}(V_i^\sigma(h_{t+1})|h_t)$ is also a submartingale.[2] To see this, recall that $Z(h_{t+1}) \geq V_i^\sigma(h_{t+1})$, so that $\mathbf{E}(Z(h_{t+1})|h_t) \geq \mathbf{E}(V_i^\sigma(h_{t+1})|h_t) = Z(h_t)$. It follows that $\mathbf{E}(V_i^\sigma(h_{t+1})|h_t)$ converges a.s. to a limit.

Finally, recalling (9.2) and writing it along any sample path for which both $V_i^\sigma(h_t)$ and $\mathbf{E}(V_i^\sigma(h_{t+1})|h_t)$ converge, we must conclude immediately that $u_i(x_t)$ converges along the very same sample path. Hence $u_i(x_t)$ converges a.s., and the equilibrium is absorbing. □

Proof of Proposition 9.2. Consider any equilibrium σ. The proof of Proposition 9.1 tells us that both $V_i^\sigma(h_t)$ and $\mathbf{E}(V_i^\sigma(h_{t+1})|h_t)$ converge to random variables V^* and $\hat{V}^*$ a.s.

By the submartingale property, $\hat{V}^* \geq V^*$ a.s., but indeed equality must hold. To see this, recall the notation $Z(h_t) \equiv \mathbf{E}(V_i^\sigma(h_{t+1})|h_t)$. Observe that $\mathbf{E}(V_i^\sigma(h_t))$ and $\mathbf{E}(Z(h_{t-1}))$ converge to $\mathbf{E}(V^*)$ and $\mathbf{E}(\hat{V}^*)$ respectively (by the dominated convergence theorem), and that $\mathbf{E}(Z(h_{t-1})) = \mathbf{E}\left[\mathbf{E}(V_i^\sigma(h_t)|h_{t-1})\right] = \mathbf{E}(V_i^\sigma(h_t))$ for every $t \geq 1$. So $\mathbf{E}(\hat{V}^*) = \mathbf{E}(V^*)$. Because $\hat{V}^* \geq V^*$ a.s., equality must hold a.s.

Consider, then, any path $\{h_t\}$ for which the above equality holds. Then the associated sequence of payoff vectors $\mathbf{u}(x_t)$, values $\mathbf{V}^\sigma(h_t)$, and conditional expectations $\mathbf{E}(\mathbf{V}^\sigma(h_{t+1})|h_t)$ *all* converge to the same limit $\mathbf{u}^*$. Because there are finitely many states, the limit of one-period payoffs is a.s. attained after finitely many dates. We claim that the same is a.s. true for $\mathbf{E}(\mathbf{V}^\sigma(h_{t+1})|h_t)$ (and trivially for $\mathbf{V}^\sigma(h_t)$ as a consequence). Suppose that the assertion is false. Then there is a positive measure of sample histories[3] such that one-period payoffs converge in finite time but the same isn't true for $\mathbf{E}(\mathbf{V}^\sigma(h_{t+1})|h_t)$. Indeed, because there are countably many dates, there is an integer S such that a positive measure of histories exists satisfying all the requirements in the preceding sentence and the additional requirement that one-shot payoffs converge by date S. Let Ω denote this distinguished set of sample paths, and let Ω^c be

---

[2] To be sure, we employ the regular version of conditional expectations in defining Z here.

[3] To be sure, this positive measure is generated by the protocol as well as equilibrium strategies.

its complement. For each path $\{h_t\} \in \Omega$ there is an individual i and a subsequence t_k such that for every k,

$$(9.4) \qquad E(V_i^\sigma(h_{t_k+1})|h_{t_k}) > u_i,$$

where u_i is the *particular* limit of one-shot payoffs for i along this path. The strict inequality in (9.4) implies that there is a another subsequence s_k of dates, with $s_k > S$ for all k, such that at each of those dates, some proposer makes a proposal which yields a higher payoff to player i than the normalized value of u_i. Because we only study pure strategies, such a proposal must be made and accepted with probability at least $\zeta > 0$, where ζ is uniform across histories and individuals, and is given by the restriction [P]. Now observe that such a proposal, if accepted, *must* subsequently lead to paths that are not in Ω. This is because (i) Ω contains only those paths for which one-period payoffs have already converged by date S, (ii) every s_k exceeds S, and (iii) an accepted proposal must lead to a change in the (by-then) stationary path of one-period payoffs. More formally,

$$\text{Prob}(\Omega^c|h_{s_k}) \geq \zeta > 0 \text{ for all } k,$$

whenever the path $\{h_t\}$ lies in Ω. It is easy to see that this must imply $\text{Prob}(\Omega^c) = 1$, a contradiction. This proves the claim that $\mathbf{u}(x_t)$, $\mathbf{V}^\sigma(h_t)$, and $E(\mathbf{V}^\sigma(h_{t+1})|h_t)$ all converge in finite time to the same limit $\mathbf{u}^*$, a.s.

We complete the proof by showing that $\mathbf{u}^*$ must be efficient. Suppose not; then there is a state x such that $\mathbf{u}(x) > \mathbf{u}^*$. Suppose a player were to propose x. The offer must be rejected, otherwise we are not in equilibrium. Consider all the rejectors: all the players who will reject conditional on all previous responders accepting. Number these players $1, \ldots, R$ in order of their appearance. For each rejector i, let h_i' denote the history following her acceptance and h_i'' the history following her rejection. Because the last rejector R rejects, it is easy to see that $V_R(h_R'') \geq u_R(x) > u_R(x^*)$. Moreover, no other player can be worse off compared to $\mathbf{u}^*$: $V_i(h_R'') \geq u_i(x^*)$ for all i. In summary,

$$(9.5) \qquad V_i(h_R'') \geq u_i(x^*) \text{ for all } i, \text{ with strict inequality for some } i.$$

Now consider player $R - 1$. She, too, rejects the offer. Therefore the first part of (9.5) holds for the history h_{R-1}''. In general, no more can be said, but because $R - 1$ is benign and (9.5) holds for the history h_R'', it must hold too for the history h_{R-1}''. Continuing recursively in this way, we see that (9.5) must holds for the history h_1''. But now we have a contradiction. By benignness, then, it is profitable for our proposer to propose x irrespective of whether it is accepted or rejected. □

9.5 Summary

This chapter studies ongoing negotiations in characteristic function games. Apart from the assumption that there are no externalities, the setup is extremely general: protocols are practically unrestricted, strategies can display history-dependence, and payoffs may or may not be transferable.

The main result of this chapter is that in characteristic functions, one-shot equilibrium payoffs must ultimately settle down, and what is more they settle down to a Pareto-efficient payoff vector.

Two important assumptions drive this result. First, we impose the condition that negotiations are, in principle, always ongoing. Second, we assume that all agents are benign, in that they do not grudge others a payoff improvement provided that they don't personally lose in the process. These two conditions are critical. Without them, counterexamples to asymptotic efficiency can easily be constructed.

With externalities across coalitions, matters are very different, and this is what we turn to next.

CHAPTER 10

Reversible Agreements With Externalities

In the last chapter, we showed that with no externalities across coalitions, a model of reversible commitments may display inefficiencies, but these are transitional. Over time, payoffs must converge to an efficient outcome. This result holds over a broad class of equilibria, including all equilibria with history-dependent strategies that satisfy a mild benignness restriction.

The purpose of this chapter is to argue that matters could be quite different when there are externalities. The ubiquitous absorption results reported for characteristic functions break down in this setting. Equilibrium payoffs may cycle, and even if they don't, inefficient outcomes may arise and persist. Finally — and in sharp contrast to characteristic functions — such outcomes are not driven by the self-fulfilling contortions of history-dependence. They occur even for Markovian equilibria.

Indeed, the ability to make sidepayments — presumably to eliminate such inefficiencies — may actually worsen the situation. In particular, the variant of our model with upfront transfers, which has thus far played a quiet and perfectly undistinguished role, now exhibits very distinctive properties.

Let's sidestep a common pitfall right away. It is tempting to think of inefficiencies as entirely "natural" equilibrium outcomes when externalities exist. Such an observation is true, of course, for games in which there are no binding agreements. Nash equilibria are "generally" inefficient, an assertion which can be given a precise formulation (see, e.g., Dubey (1986)). When agreements can be costlessly written, however, no such presumption can and should be entertained. These are models of *binding* agreements, a world in which the so-called "Coase theorem" is relevant. For instance, in all that we've done so far, two-player games invariably yield efficiency, quite irrespective of whether there are externalities across the two players. This is not to say that the "usual intuition" has no role to play in the events of this chapter. It must, because the process of negotiation is itself modeled as a noncooperative game. But that is a very different object from the "stage game" over which agreements are sought to be written.

For the material in this chapter, I continue to rely on Gomes and Jehiel (2005) and Hyndman and Ray (2007). We begin with the baseline model and then move on to the variant with upfront transfers.

10.1 The Baseline Model for Three-Player Games

Three-player games represent an interesting special case. Even when externalities are allowed for, such games share a central feature with their characteristic function counterparts: each player possesses, in effect, a high degree of veto power in all moves that alter her payoff. This will allow us to prove a limited efficiency result, even when externalities are widespread.

It is worth noting that three-player situations have been the focus of study in several applied models of coalition formation (see, e.g., Krishna (1998), Aghion, Antras and Helpman (2004), Kalandrakis (2004) and Seidmann (2005)).

10.1.1 The Failed Partnership. Begin with an example. Suppose that there are three agents, any two of whom can become "partners". For instance, two of three countries could form a customs union, or a pair of firms could set up a production cartel or an R&D coalition with a commitment to share ideas. I presume that the outsider to the partnership gets a "low" payoff: zero, say. Finally, a three-player partnership is assumed not to be feasible (or has very low payoffs).

The crucial feature of this example is that player 1 is a bad partner, or — for the purposes of better interpretation — a *failed partner*. Partnerships between him and any other individual are dominated — both for the partners themselves and for the outsider — by all three standing alone. In contrast, the partnership between agents 2 and 3 is rewarding (for those agents).

We formalize this as a partition function game. In the examples that follow, we simply record those states with nontrivial payoff vectors, and omit any mention of the remaining states, with the presumption that the payoffs in those states are zero to all concerned. We shall also be somewhat cavalier in our description of equilibrium and ignore these trivial states: equilibrium transitions from those states are implicitly defined in obvious ways.

EXAMPLE 10.1. *Consider the following three-player game with minimal approval committees:*

$$\begin{aligned}
x_0 &: \pi_0 = \{\{1\}, \{2\}, \{3\}\}, & \mathbf{u}(x_0) &= (6,6,6) \\
x_1 &: \pi_1 = \{\{1\}, \{23\}\}, & \mathbf{u}(x_1) &= (0,10,10) \\
x_2 &: \pi_2 = \{\{2\}, \{13\}\}, & \mathbf{u}(x_2) &= (5,0,5) \\
x_3 &: \pi_3 = \{\{12\}, \{3\}\}, & \mathbf{u}(x_3) &= (5,5,0).
\end{aligned}$$

OBSERVATION 10.1. *For δ sufficiently close to 1 in Example 10.1, the outcomes x_2 and x_3 — which are inefficient — must be absorbing states in every equilibrium.*

A formal proof of this observation isn't needed; the discussion to follow will suffice. Why might x_2 and x_3 be absorbing? The reason is very simple. Despite the fact that x_2 (or x_3) is Pareto-dominated by x_0, player 1 won't accept a transition to x_0. If she did, players 2 and 3 would initiate a further transition to x_1. Player 1 *might* accept such a transition if she is very myopic and prefers the short-term payoff offered by x_0, but if she is patient enough she will see ahead to the infinite phase of "outsidership" that will surely follow the short-term gain. In that situation it will be impossible to negotiate one's way out of x_2 or x_3. This inefficiency persists in *all* equilibria, history-dependent or otherwise.

Notice that x_2 or x_3 wouldn't be *reached* starting from any other state. This is why the interpretation, the "failed partnership", is useful. The example makes sense in a situation in which players have been locked in with 1 on a past deal, on expectations which have failed since. To be sure, this interpretation is unnecessary for the formal demonstration of persistent inefficiency from some initial state.

Notice that the players *could* negotiate themselves out of x_2 if 2 and 3 could credibly agree never to write an agreement while at x_0. Are such promises reasonable in their credibility? One could certainly assume that they are (economists and game theorists have been known to assume worse). However, it may be difficult to imagine that from a legal point of view, player 1, who has voluntarily relinquished all other contractual agreements between 2 and 3, could actually hold 2 and 3 to such a meta-agreement.

This example raises three important points. The first is an immediate outgrowth of the previous discussion. Does one interpret the stand-alone option (x_0) as an *agreement* from which further deviations require universal permission? Or does "stand-alone" mean freedom from all formal agreement, in which case further bilateral deals only need the consent of the two parties involved? Our discussion takes the latter view.

Second, observe that the lack of superadditivity in this example is important. If the grand coalition can realize the Pareto-improvement then player 1 can control any subsequent shenanigans by 2 and 3, and he will therefore permit the improvement. The issue of superadditivity is one to which we shall return below.

Finally, recall that upfront transfers are not permitted in this example. Were they allowed in unlimited measure, players 2 and 3 could reimburse player 1 for the present discounted value of his losses in relinquishing his partner. Depending on the discount factor, the amounts involved may be considerable. But they would break the deadlock. But upfront transfers have other, more subtle implications, and here too we must postpone the discussion to a later stage.

10.1.1.1 An Efficiency Result for Three-Person Games. The failed partnership or its later variant is not the only form of inefficiency that can arise. Appendix A to this chapter records three other forms of inefficiency, including one which can even arise from the stand-alone starting point of no agreements: the structure of singletons. In the light of these several examples, it is perhaps of interest that a positive (though limited) *efficiency* result holds for every three-person game satisfying a "minimal transferability" restriction. To state this restriction, let $\bar{u}(i, \pi)$ be the maximum one-period payoff to player i over all states with the same coalition structure π.

[T] If two players i and j both belong to the same coalition in coalition structure π, then $\bar{u}(i, \pi)$ and $\bar{u}(j, \pi)$ are achieved at different states.

PROPOSITION 10.1. *Consider a three-person game with a finite number of states and satisfying condition T, with history-independent proposer protocols and minimal approval committees. Then for all δ close enough to 1, there exists an initial state and a stationary Markov equilibrium with efficient absorbing payoff limit from that state.*

The proof of Proposition 10.1 exhaustively studies different cases, and is therefore relegated to Appendix B.[1] But we can provide some broad intuition for the result. Pick any player i and consider her maximum payoff over all conceivable states. If this maximum is attained at a state x^* in which i belongs to a coalition with two or more players, then observe that i's consent *must* be given for the state to change. (This step is not true when there are four or more players, and invalidates the proposition, as we shall see later.) Because the payoff in question is i's maximum, it is easy enough to construct an equilibrium in which x^* is an absorbing state.

It therefore remains to consider games in which for every player, the maximum payoff is attained at states in which that player stands alone. If no such state is absorbing in an equilibrium, one can establish the existence of a cyclical equilibrium path, the equilibrium payoffs along which are uniquely pinned down by the payoffs at the state in which all players stand alone. With the transferability condition T, one can now find payoff vectors for other coalitions (doubletons or more) such that *some* player in those coalitions prefer these payoffs to the cyclical equilibrium payoffs. The associated states then become absorbing, and a simple additional step establishes their efficiency.

We conjecture that neither the minimality of approval committees nor the history-independence of proposer protocols is needed for this result, but do not have a proof.

10.2 The Baseline Model for Four or More Players

The analysis in the previous section shows that once externalities are introduced, a failure of efficiency is a distinct possibility. In

---

[1] It should be noted that in most cases the result is stronger in that it does not insist upon $\delta \to 1$; in only one case do we rely on $\delta \to 1$.

the example of the failed partnership, one agent holds his partner hostage in the fear that if the partner is relinquished (so as to create a Pareto improvement), other deals will subsequently be struck that leave our agent with very low payoffs.

Yet, as Proposition 10.1 goes on to show, it isn't possible to create this phenomenon from *every* initial state. In the example of the failed partnership, the very state that the failed partner fears is undeniably Pareto-efficient. It's true that the failed partner suffers in this state, but the other two agents are certainly as well off as they can be. If negotiations were to commence with *this* state as initial condition, the resulting outcome must be efficient. In short, if one state isn't efficient something else must be, and we must ultimately arrive at some such state that's absorbing. This is the content of Proposition 10.1.

But don't make the mistake of supposing that such a proposition must follow from a simple process of eliminating inefficient states. Indeed, the proposition fails when the number of players exceeds three. To show this, I present an example that displays the most severe form of inefficiency: *every* absorbing state in every Markovian equilibrium is static inefficient, and every nonconvergent equilibrium path in every Markovian equilibrium is dynamically inefficient.

EXAMPLE 10.2. *Consider the following four-player game with minimal approval committees:*

$$x_1 : \pi_1 = \{\{12\}, \{3\}, \{4\}\}, \quad \mathbf{u}(x_1) = (4, 4, 4, 4)$$
$$x_2 : \pi_2 = \{\{1\}, \{2\}, \{3\}, \{4\}\}, \quad \mathbf{u}(x_2) = (5, 5, 5, 5)$$
$$x_3 : \pi_3 = \{\{1\}, \{2\}, \{34\}\}, \quad \mathbf{u}(x_3) = (0, 0, 10, 10)$$
$$x_4 : \pi_4 = \{\{12\}, \{34\}\}, \quad \mathbf{u}(x_4) = (2, 2, 2, 2).$$

Assume that a fresh proposer is chosen with uniform probability at each proposal stage.

OBSERVATION 10.2. *For δ sufficiently close to 1 in Example 10.2, every stationary Markov equilibrium is inefficient starting from any initial state.*

The proof of this result may be found in Appendix C.

As we've already discussed, the fact that *some* absorbing state in *some* equilibrium may be Pareto-dominated is not too surprising. In part, a similar logic is at work here. Begin with the state x_1, in which players 1 and 2 are partners and 3 and 4 are separate. This is a failed partnership (at least in a context in which players 3 and

4 are *not* partners themselves): if the {12}-partnership disbands, the state moves to x_2 which is better for all concerned. But once at x_2, we see other latent, beneficial aspects of the erstwhile partnership between 1 and 2: if players 3 and 4 now form a coalition, they can exploit 1 and 2 for their own gain; this is the state x_3.

So far, the story isn't too different from that of the failed partnership. But the similarity ends as we take up the story from the point at which 3 and 4 fashion their counterdeviation to x_3. Their gains can be reversed if players 1 and 2 form (or depending on the dynamics, re-form) a coalition. Balance is now restored; this is the state x_4.[2] Finally, in this context, the partnership between 3 and 4 is more a hindrance than a help (just as {12} was in the state x_1), and they have an incentive to disband. We are then "back" to x_1.

This reasoning appears circular, and indeed in a sense it is, but such a circularity is in fact the essential content of Observation 10.2: as long as equilibria are Markovian, there is asymptotic inefficiency from *every* initial condition, despite the ability to write and renegotiate permanently binding agreements.

One might suspect that the Observation is vacuous in that no Markov equilibrium, efficient or not, exists. I could allay such suspicions by appealing to the existence theorem mentioned in Chapter 8, but it may be better to display such equilibria explicitly. Here is one. In it:

State x_1 is absorbing.

State x_2 moves back to x_1 when 1 or 2 propose, and on to x_3 whenever 3 or 4 propose.

State x_3 moves to x_4 when 1 or 2 propose, and remains unchanged otherwise.

State x_4 moves to x_1 no matter who proposes.

Observe that the state x_1, which is plainly Pareto-dominated, is not just absorbing but "globally" absorbing.[3]

To verify that this description constitutes an equilibrium, begin with state x_1. Obviously players 3 and 4 do not benefit from changing the

---

[2] So the partnership {12} is not entirely a failure; it depends on the context.
[3] We are neglecting the trivial states with zero payoffs for all. Including them would obviously make no difference.

state to x_4, which is all they can unilaterally do. Players 1 and 2 can (bilaterally) change the state to x_2, by the presumed minimality of approval committees. If they do so, the subsequent trajectory will involve a stochastic path back to x_1.[4] Some fairly obvious but tedious algebra reveals that the Markov value function $V_i(x, \delta)$ satisfies

$$V_i(x_2) = 5 - 3\delta + \frac{\delta^2(1 + \delta)}{2\left(1 - \frac{\delta}{2}\right)}$$

for $i = 1, 2$. This value converges (as it must) to that of the absorbing state — 4 — as delta goes to 1, but the important point is that the convergence occurs "from below", which means that $V_i(x_2, \delta)$ is strictly smaller than $V_i(x_1, \delta) = 4$ for all delta close enough to 1.[5] Perhaps more intuitively but certainly less precisely, the move to state x_2 starts off a stochastic cycle through the payoffs 5, 0 and 2 before returning to absorption at 4, which is inferior to being at 4 throughout. This verifies that players 1 and 2 will relinquish the opportunity at x_1 to switch the state to x_2. It also proves that once at state x_2, players 1 and 2 will want to return to the safety of x_1 if they get a chance to move.

On the other hand, players 3 and 4 will want to move the state from x_2 to x_3. Proving this requires more value-function calculation. A second round of tedious algebra reveals that

$$V_i(x_3, \delta) - V_i(x_2, \delta) = 5 - 6\delta + \delta^2$$

for $i = 3, 4$. This difference vanishes (as it must) as δ approaches 1, but once again the important point is that the difference is strictly positive for all δ close to 1 (indeed, for all δ), which justifies the move of 3 and 4.

That 1 and 2 must want to move away as quickly as possible from state x_3, and 3 and 4 not at all, is self-evident. That leaves x_4. At this state players 3 and 4 receive their worst payoffs, and will surely

---

[4]We are arguing in the spirit of the one-shot deviation principle, in which the putative equilibrium strategies are subsequently followed. Even though the one-shot deviation principle needs to be applied with care when coalitions are involved, there are no such dangers here as all coalitional members have common payoffs.

[5]We verify this by differentiating $V_i(x_2, \delta)$ with respect to δ and evaluating the derivative at $\delta = 1$.

want to move to x_1, and indeed, players 1 and 2 will want that as well.[6] Our verification is complete.

There is actually a second equilibrium with no absorbing states, in which players 1 and 2 randomize between states x_1 and x_2, while players 3 and 4 randomize between states x_3 and x_4. While we omit the details, it is easy to check that such an equilibrium displays (dynamic) inefficiency from every initial condition, because it *must* spend nonnegligible time at the inefficient states x_1 and x_4. We omit the details.

Two final remarks are in order regarding this example. First, the strong form of inefficiency is robust to (at least) a small amount of transferability in payoffs. The reason is simple; given the payoffs at x_3 (resp. x_4), the payoffs to players 1 and 2 (resp. 3 and 4) are still minimal. Therefore, these states *cannot* be absorbing. But then, even with a little transferability, we are in the same situation as in the example.

Second, the example is not robust to the use of history dependent strategies. Indeed, x_2 can be supported as an absorbing state provided that deviations from x_2 are punished by a return to the inefficient stationary equilibrium in which x_1 is absorbing.

This last remark creates an interesting contrast between models based on characteristic functions and those based on partition functions. In the former class of models, the work of Seidmann and Winter (1988) and Okada (2000) assure us that ongoing negotiations lead to efficiency under Markovian equilibrium. It's the possibility of history-dependence that creates the inefficiency problem, albeit one that we successfully resolved in Chapter 9 with the help of the benignness condition. In contrast, partition functions are prone to inefficiency under Markovian equilibrium, as Examples 10.1 and 10.2 illustrate. History-dependence might help to alleviate this problem (it does in Example 10.2, though not in Example 10.1).

---

[6]Because we've developed the state space model at some degree of abstraction, we've allowed any player to make a proposal to any coalition, whether or not she is a member of that coalition. This is why players 1 and 2 ask 3 and 4 to move along. Nothing of qualitative import hinges on allowing or disallowing this feature. The transition from x_4 back to x_1 would still happen, but more slowly.

10.3 Superadditive Games

An important feature of the examples in Section 10.2 is that they employ a subadditive payoff structure. Is that a reasonable assumption? This is a subtle question that recalls the discussion in Section 3.4 of Chapter 3. We continue that discussion here.

First, it should be noted that in games with externalities superadditivity is generally not to be expected. For instance, recall the example of the Cournot oligopoly studied in detail in Chapter 6. Using the partition function developed there, it is easy to see that if there are just three firms, firms 1 and 2 do worse together than apart, provided that firm 3 stands separately in both cases.

At the same time, this argument does not apply to the *grand* coalition of all firms. Indeed, every partition function derived from a game in strategic form must satisfy *grand coalition superadditivity* (GCS):

[GCS] For every state $x = (u, \pi)$, there is $x' = (u', \{N\})$ such that $u' \geq u$.

Is GCS a reasonable assumption? In Chapter 3 we've argued that in many cases it may not be. To continue that discussion without undue repetition, one possible interpretation of GCS is that it is a "physical" phenomenon; e.g., larger groups organizing transactions more efficiently, or sharing the fixed costs of public good provision. Yet such superadditivities are often the exception rather than the rule. After all, the entire doctrine of healthy competition is based on the notion that physical superadditivity, after a point, is not to be had. In general, too many cooks do spoil the broth: competition among groups can lead to efficiency gains not possible when there is a single, and perhaps larger, group attempting to act cooperatively. In addition to competition, Section 3.4 lists a host of other reasons for lack of physical superadditivity.[7]

But some game theorists might argue that this isn't what is meant by superadditivity at all. They have in mind a different notion of GCS, which is summarized in the notion of the *superadditive cover*. After all, the grand coalition can write a contract which

---

[7]In all of the cases, the argument must be based on some noncontractible factor, such as the creativity or productivity created by the competitive urge, or ideological differences, or the presence of stand-alone players who are outside the definition of our set of players but nevertheless have an effect on their payoffs.

exactly replicates the payoffs obtainable in some other coalition structure. For instance, companies do spin off certain divisions, and organizations do set up competing R&D groups. In a word, the grand coalition can agree not to cooperate, if need be.

In a static setting, such a position represents, perhaps, no loss of generality. But in a dynamic setting this view embodies a crucial assumption: that future changes in the strategy of (or in the alliances formed by) one of the subgroups will require the consent of the entire grand coalition of which that group was supposedly a part. For example, consider contracts between senior executives and firms. They typically contain a clause enjoining the executive from working for a competitor firm for a number of years — so-called *no compete* clauses. To some extent, this reflects the notion of the superadditive cover: surely, if all parties agreed, the executive would be free to work for the competitor, while if the original firm dissents, she would not — at least for a certain length of time. To the extent that such contracts *cannot* be enforced for an infinite duration, the model without grand-coalition superadditivity can be viewed as a simplification of this, and other, real-world situations.

Nevertheless, GCS applies without reservation to many other cases. So it is worth recording that GCS restores (Markovian) efficiency, at least if the existence of an absorbing limit payoff is assumed:

PROPOSITION 10.2. *Under GCS, every absorbing payoff limit of every Markovian equilibrium must be static efficient.*

The proof follows a much simpler version of the argument for Proposition 9.2 and we omit it.

It must be noted, however, that GCS does not guarantee long-run efficiency in all situations: Example 10.2 can be modified so that GCS holds *but* there is an inefficient cycle over states that do not involve the grand coalition. In order to guarantee that even this form of inefficiency does not persist in the long-run, one needs enough transferability of payoffs within the grand coalition. Indeed, it can be proved that under GCS and the additional assumption that the payoff frontier for the grand coalition is continuous and concave, *every* Markov equilibrium must be absorbing — and therefore asymptotically efficient.

10.4 Upfront Transfers and the Failure of Efficiency

How does the ability to make transfers affect the examples? It is important to distinguish between two kinds of transfers. Coalitional or partnership worth could be freely transferred between the players within a coalition. Additionally, players might be able to make large upfront payments in order to induce certain coalitions to form. In all cases, of course, the definition of efficiency should match the transfer environment.[8]

Within-coalition transferability often does nothing to remove inefficiency. For instance, nothing changes in the failed partnership of Example 10.1. On the other hand, upfront transfers *across* coalitions have an immediate and salubrious effect in that example. Efficiency is restored from every initial state. The reason is simple. If player 1 is offered any (discount-normalized) amount in excess of 5, he will "release" player 2. In view of the large payoffs that players 2 and 3 enjoy at state x_1, they will be only too pleased to make such a payment. The final outcome, then, from any initial condition is the state x_1, and we have asymptotic efficiency. It is true that the amount of the transfer may have to be enormous when the discount factor is close to 1, but we've already discussed that (see Section 8.2.2 in Chapter 8). Our concern here is with the *implications* of the upfront-transfer scenario.

The beauty of the Gomes–Jehiel (2005) paper, on which I now proceed to rely, is that it unearths an entirely different face of upfront transfers. To see this, I introduce a seemingly innocuous variant of Example 10.1:

EXAMPLE 10.3. *Consider the following three-player game with minimal approval committees:*

$$x_0 : \pi_0 = \{\{1\}, \{2\}, \{3\}\}, \ \mathbf{u}(x_0) = (6, 6, 1)$$
$$x_1 : \pi_1 = \{\{1\}, \{23\}\}, \quad \mathbf{u}(x_1) = (0, 7, 2)$$
$$x_3 : \pi_3 = \{\{12\}, \{3\}\}, \quad \mathbf{u}(x_3) = (5, 5, 0)$$

and all other states have payoff 0.

---

[8]For instance, if transfers are not permitted, it would be inappropriate to demand efficiency in the sense of aggregate surplus maximization. If an NTU game displays inefficiency in the sense that "aggregate surplus" is not maximized, this is of little interest: aggregate surplus is simply the wrong criterion.

First, take a quick look at this example without introducing upfront transfers and compare it with its predecessor, Example 10.1. Little of substance has been changed, though we've broken symmetry by assigning zero to any partnership between Players 1 and 3. Players 1 and 2 still form a "failed partnership" at x_3, yet at that state player 1 continues to cling to player 2. They would prefer to move to state x_1, but player 1 rationally fears the subsequent switch to x_2, something that is out of his control.

But the introduction of upfront transfers in this example has a perverse effect. Instead of taking the inefficiency away (as it does in Example 10.1), it generates inefficiency from *every* initial condition.

OBSERVATION 10.3. *In Example 10.3, every stationary Markov equilibrium is inefficient starting from any initial state.*

It isn't hard to see what drives the assertion. With unlimited upfront transfers, we are entitled to add payoffs across players to derive our efficiency criterion. Only the state x_0 is efficient on this score, and if an equilibrium is to display asymptotic efficiency — whether dynamic or static — it can stay away from x_0 for a finite number of dates at best. Now the root of the trouble is clear: players 2 and 3 invariably have the incentive to move away from x_0 to x_1. Not that matters will come to an end there: the fact that x_1 is itself inefficient will cause further movement across states as upfront transfers continue to be made along an infinite subsequence of time periods. The precise computation of such transfers is delicate,[9] but the assertion that efficiency cannot be attained should be clear.

Actually, Example 10.3, while it makes the point well enough, obscures a matter of some interest. In that example, players 2 and 3 gain on two counts when they move the state from x_0 to x_1. They make an immediate gain in payoffs, and then they gain even more subsequently as they are paid additional ransom in the form of upfront transfers. The point of the next example is that the "ransom effect" dominates, at least when discount factors are close to 1.

---

[9]Such transfers will have to be made with a rational eye on the fact that an endless cycle across states will, in fact, happen.

EXAMPLE 10.4. *Consider the following three-player game with minimal approval committees:*

$$x_0 : \pi_0 = \{\{1\},\{2\},\{3\}\}, \quad \mathbf{u}(x_0) = (b,b,b)$$
$$x_1 : \pi_1 = \{\{1\},\{23\}\}, \quad \mathbf{u}(x_1) = (0,a,a)$$
$$x_2 : \pi_2 = \{\{2\},\{13\}\}, \quad \mathbf{u}(x_1) = (a,0,a)$$
$$x_3 : \pi_3 = \{\{12\},\{3\}\}, \quad \mathbf{u}(x_3) = (a,a,0)$$

and all other states have payoff zero. Assume that $b > a > 0$.

This example has none of the asymmetries of Example 10.3. There is a unique efficient state by any criterion. It is state x_0. It Pareto-dominates every other state. Players moving away from this state suffer an immediate and unambiguous loss in payoffs. Yet we have

OBSERVATION 10.4. *In Example 10.4, every stationary Markov equilibrium under the uniform proposer protocol is inefficient starting from any initial state: the state x_0 can never be absorbing.*

This observation highlights very cleanly the negative effects of upfront transfers. Players may deliberately generate inefficient outcomes to seek such transfers as ransom. This effect is particularly clear in Example 10.4 because each of the states x_1, x_2 and x_3 are Pareto-inferior to x_0.

It will be instructive to work through this example by informally proving Observation 10.4. For simplicity, we will spell out the details for all *symmetric* Markovian equilibria.

Suppose, contrary to the claim, that x_0 is absorbing. Then the (discount-normalized) value to each player is just b: $V_i(x_0) = b$ for all i. In each of the other states x_i, player i is an "outsider" currently earning 0; denote her lifetime payoff by c, evaluated *ex ante*, before a proposer has been determined. The other two players are partners, denote by d their corresponding payoffs.

It is very easy to see that d is at least as large as c: the partners can make all the acceptable proposals that the outsider can make, while enjoying a current payoff which exceeds that of the outsider. Moreover, because x_0 is absorbing, it must be that $b \geq d$, otherwise some pair would surely destabilize x_0.

The last of our preliminary observations is that no proposer in any of the nonabsorbing states x_i stands to gain anything by switching the state to another nonabsorbing state x_j: the sum of all payoffs is constant (at $2d + c$) so there is no surplus to be grasped. On the other hand, both the partners and the outsiders make a gain by steering the state back to x_0. Consequently, recalling that proposer protocol is uniform, we have

$$(10.1) \quad d = \frac{1}{3}[b + \{b - (1 - \delta)a - \delta d\} + (b - \delta c)] + \frac{2}{3}[(1 - \delta)a + \delta d].$$

The reason is simple. Take any partner at one of the nonabsorbing states. With probability 1/3, she gets to propose. She successfully moves the state straight away to x_0, earning a lifetime payoff of b, *and* can demand an upfront transfer of up to $b - (1 - \delta)a - \delta d$ from her partner, and up to $b - \delta c$ from the outsider.[10] With remaining probability 2/3, she is proposed *to*, in which case she will be driven down to her reservation value, which is precisely $(1 - \delta)a + \delta d$. Together, this gives us (10.1).

A parallel argument tells us that as far as the outsider is concerned,

$$(10.2) \qquad c = \frac{1}{3}[b + 2\{b - (1 - \delta)a - \delta d\}] + \frac{2}{3}\delta c.$$

The reasoning is very similar to that underlying (10.1), and we'll skip the repetition.

Now combine (10.1) and (10.2) and simplify to see that

$$d = b + \frac{a}{3},$$

which contradicts our initial presumption that $b \geq d$. We have therefore proved that the unique efficient outcome cannot be stable. Consequently, the equilibrium path, no matter what it looks like, must display persistent inefficiency.

Notice that (in contrast to Example 10.3), the deviating players do suffer a loss in current payoff when they move away from the efficient state. But the prospect of inflicting a *still* greater loss on the outsider raises the possibility that the outsider will pay to have the state moved back — albeit temporarily — to the efficient point. This is a new angle on upfront transfers. They may lubricate the path to efficiency, but they might encourage deviations from efficient paths as well, in order to secure a ransom. Thus the presumption that unlimited transfers act to restore or maintain efficiency is wrong.

---

[10]If her partner refuses, she enjoys $(1 - \delta)a$ today and starting tomorrow, a present discounted value of δd. She will therefore agree to any proposal that gives her more than this amount. A similar argument holds for the outsider, whose payoff conditional on refusal is just δc.

One more feature of Example 10.4 is worth mentioning. The efficient state is one in which all three players stand apart. This is precisely what makes that state persistently unstable, for two players can always form an inefficient coalition. If this contractual right can be eliminated in the act of making an upfront transfer, then efficiency can be restored: once state x_0 is regained, there can be no further deviations from it. This line of discussion is exactly the same as in Section 10.3, and there is nothing further to add here.

More generally, the efficient state in this example has the property that a subset of agents can move away from that state (i.e., they can act as approval committee for a move) such that some *other* agents — not on the approval committee — are thereby rendered worse off in terms of current payoffs. Whenever this is possible, there is scope for collecting a ransom, and the potential for a breakdown in efficiency.Gomes and Jehiel (2005) develop this idea further.

10.5 Summary

Our study of coalitional bargaining problems in "real time" yields a number of implications. For characteristic function form games, a very general result for all pure-strategy equilibria (whether history-dependent or not) can be established: every equilibrium path of states must eventually converge to some absorbing state, and this absorbing state must be static efficient. This was the subject of Chapter 9.

In contrast, in games with externalities, matters are more complicated and none of the results for characteristic function games continue to hold without further conditions. It is easy enough to find a three-person example in which there is persistent inefficiency from some initial state, whether or not equilibria are allowed to be history-dependent. At the same time, we also show that in every three-person game, there is *some* Markovian equilibrium which yields asymptotic efficiency from *some* initial condition.

Yet even this limited efficiency result is not to be had in four-person games. Section 10.2 demonstrates the existence of games in which every absorbing state in every Markovian equilibrium exhibits asymptotic static inefficiency.

The situation is somewhat alleviated if the game in question exhibits grand coalition superadditivity. In that case, it is possible to recover efficiency, provided that the equilibrium is absorbing.

Finally, we show that the ability to make unlimited upfront transfers may worsen the efficiency problem.

The main open question for games with externalities is whether there always exists some history-dependent equilibrium which permits the attainment of asymptotic efficiency from *some* initial state (that there is no hope in obtaining efficiency from *every* initial state is made clear in Section 10.1.1). I am pretty sure that the answer should be in the affirmative: assuming — by way of contradiction — that equilibria are inefficient from every initial state, one should be able to employ such equilibria as continuation punishments in the construction of some efficient strategy profile. Such a result would be intuitive: after all, one role of history-dependent strategies is to restore efficiency when simpler strategy profiles fail to do so.

Finally, the general setup in Section 8.2 may be worthy of study, with or without binding agreements. For instance, the general setup is applicable to games in which agreements are only temporarily binding, or in which unanimity is not required in the implementation of a proposal. There is merit in exploring these applications in future work.

Appendix A

Other Examples For Three-Player Games.

In the examples below, if a coalition structure is omitted, it means that either every player obtains an arbitrarily large negative payoff or there is some legal impediment to the formation of that coalition structure. In all of the examples of this section, we assume minimal approval committees; for example, from the singletons, players 1 and 2 can approve a transition to any state y with coalition structure $\{\{12\}, \{3\}\}$.

More on Inefficiency. One response to the inefficiency example of Section 4.1.1 in the main text is that the inefficient state described there will never be reached *starting from the singletons*. Setting the initial state to the singletons has special meaning: presumably this is the state from which all negotiations commence. However, this is wrong on two fronts (at least for Markov equilibria) as we now show.

Coordination Failures. Coordination failures, leading to inefficiency from *every* initial state, are a distinct possibility, even in three player games. Consider the following:

$$x_0 : \pi_0 = \{\{1\}, \{2\}, \{3\}\}, \; \mathbf{u}(x_0) = (2, 2, 2)$$
$$x_1 : \pi_1 = \{\{1\}, \{23\}\}, \; \mathbf{u}(x_1) = (-1, 1, 1)$$
$$x_2 : \pi_2 = \{\{2\}, \{13\}\}, \; \mathbf{u}(x_2) = (1, -1, 1)$$
$$x_3 : \pi_3 = \{\{12\}, \{3\}\}, \; \mathbf{u}(x_3) = (1, 1, -1).$$

OBSERVATION 10.5. *Suppose that everyone proposes with equal probability at every date. Then, for $\delta \in [\frac{3}{5}, 1)$, there is an MPE in which x_i is absorbing, and from x_0, there is a transition to x_i with probability $\frac{1}{3}$ for $i = 1, 2, 3$.*

The proof is simple and we omit it.

Convergence to Inefficiency From The Singletons. Consider the following example, which is a variation on the "failed partnership" example of Section 4.1.1.

$$x_0 : \pi_0 = \{\{1\}, \{2\}, \{3\}\}, \; \mathbf{u}(x_0) = (5, 5, 5)$$
$$x_1 : \pi_1 = \{\{1\}, \{23\}\}, \; \mathbf{u}(x_1) = (0, 6, 8)$$
$$x_2 : \pi_2 = \{\{2\}, \{13\}\}, \; \mathbf{u}(x_2) = (3, 0, 10)$$
$$x_3 : \pi_3 = \{\{12\}, \{3\}\}, \; \mathbf{u}(x_3) = (4, 4, 0).$$

OBSERVATION 10.6. *For any history-independent proposer protocol such that at x_0 each player has strictly positive probability of proposing, there exists $\bar{\delta} \in (0, 1)$ such that if $\delta \geq \bar{\delta}$, all stationary Markovian equilibria involve a transition from x_0 to x_3 — and full absorption into x_3 thereafter — with strictly positive probability.*

Proof. Let $\alpha = (\alpha_1, \alpha_2, \alpha_3) \in \text{int}(\Delta)$ denote the proposers' protocol at x_0. First notice that in every equilibrium x_1 and x_2 must be absorbing. The states x_1 and x_2 give players 2 and 3, respectively, their unique maximal payoff. Moreover, at x_1 (resp. x_2) player 2 (resp. player 3) has veto power over any transition. Second, in every equilibrium, x_0 cannot be absorbing. This follows because players 2 and 3 can always initiate a transition to x_1 and earn a higher payoff.

We now proceed with the rest of the proof. First, we rule out a "cycle" by proving the following: If there is a positive probability transition from x_0 to x_3, then x_3 must be absorbing. Indeed, suppose not. Then for $i = 1, 2$, $V_i(x_0) = V_i(x_3) = 4$. But then, from x_0, player 1 will always reject a transition to x_2, which means that $V_2(x_0) \geq 5$, a contradiction.

Next suppose that the probability of reaching x_3 from the singletons is zero. Observe that $V_1(x_0) \leq 3$, for if not, x_1 is the only absorbing state reachable from x_0, implying that $V_1(x_0) \to 0$ for δ sufficiently high, a contradiction. Similarly, $V_3(x_0) \leq 8$, for if not, x_2 is the only absorbing state reachable from the singletons. But then for δ sufficiently high, $V_2(x_0) \leq 4$, implying that players 1 and 2 would initiate a transition to x_3, a contradiction. Finally, observe that since x_3 is not reached with positive probability, it must be that $V_2(x_0) \geq 4$, since otherwise, 1 would offer x_3 and it would be accepted.

Let p_i denote the probability of a transition from x_0 to x_i for $i = 0, 1, 2$. By assumption, $p_3 = 0$ and we have just shown that $p_1, p_2 > 0$. Given p_i, write the equilibrium value functions and take the limit as $\delta \to 1$ to obtain:

$$\overline{V}_1(x_0) = \frac{3p_2}{1-p_0} \leq 3$$

(10.3)
$$\overline{V}_2(x_0) = \frac{6p_1}{1-p_0} \geq 4$$

$$\overline{V}_3(x_0) = \frac{8p_1 + 10p_2}{1-p_0} \leq 8.$$

From the third equation in (10.3), we see that $p_2 = 0$, which then implies that the first equation is satisfied with strict inequality. Therefore, player 1 strictly prefers to propose x_2, and the offer will be accepted by player 3. Hence, $p_2 > \alpha_1 > 0$, a contradiction. It then follows that for δ sufficiently high the same conclusion may be drawn. □

Cyclical Equilibria. Next, equilibrium *cycles* become a distinct possibility:

$$x_0 : \pi_0 = \{\{1\}, \{2\}, \{3\}\}, \quad \mathbf{u}(x_0) = (1, 1, 1)$$
$$x_1 : \pi_1 = \{\{1\}, \{23\}\}, \quad \mathbf{u}(x_1) = (0, 2, 1)$$
$$x_2 : \pi_2 = \{\{2\}, \{13\}\}, \quad \mathbf{u}(x_2) = (\tfrac{1}{2}, 4, 1).$$

OBSERVATION 10.7. *Suppose that everyone proposes with equal probability at every date. Then there is an equilibrium with the following transitions:*

$$x_0 \underset{2/3}{\to} x_1 \underset{1}{\to} x_2 \underset{2/3}{\to} x_0.$$

Dynamic Inefficiency In Every Equilibrium. Though we did not formally prove this for characteristic functions, every Markovian equilibrium must exhibit full *dynamic* efficiency from *some* initial state. This is no longer true for games with externalities:

$$x_0 : \pi_0 = \{\{1\}, \{2\}, \{3\}\}, \quad \mathbf{u}(x_0) = (1, 1, 1)$$
$$x_1 : \pi_1 = \{\{1\}, \{23\}\}, \quad \mathbf{u}(x_1) = (10, 0, 0)$$
$$x_2 : \pi_2 = \{\{2\}, \{13\}\}, \quad \mathbf{u}(x_2) = (0, 10, 0)$$
$$x_3 : \pi_3 = \{\{12\}, \{3\}\}, \quad \mathbf{u}(x_3) = (0, 0, 10).$$

If x_i, $i = 1, 2, 3$ were absorbing, then for $j \neq i$, $V_j(x_i) = 0$. However, notice that in every Markovian equilibrium, for all $i = 1, 2, 3$, $V_i(x_0) \geq 1$. Therefore, j *must* accept a proposal from x_i to x_0, hence a profitable deviation exists. Finally, it can be shown that any cyclical Markovian equilibrium must necessarily spend time at x_0. We have therefore proved:

OBSERVATION 10.8. *Suppose that everyone proposes with equal probability at every date. Then every Markovian equilibrium exhibits dynamic inefficiency from every initial state.*

Appendix B

Proof of Proposition 10.1.

In what follows, we denote by π_0 a singleton coalition structure, by π_i a coalition structure of the form $\{\{i\}, \{j, k\}\}$, and by π_G the structure consisting of the grand coalition alone. Use the notation $\pi(x)$ for the coalition structure at state x and $S^i(x)$ for the coalition to which i belongs at x. Subscripts will also be attached to states (e.g., x_i) to indicate the coalition structure associated with them (e.g., $\pi(x_i) = \pi_i$).

For each i, let $X_i^* = \text{argmax}\{u_i(x) \mid x \in X\}$, with x_i^* a generic element. Finally, we will refer to $\pi(x_i^*) = \pi$ as a *maximizing* (coalition) structure (for i).

Case 1: There exists $i = 1, 2, 3$ and $x_i^* \in X_i^*$ such that $|S^i(x_i^*)| \geq 2$.

Pick $x_i^* \in X_i^*$ as described and consider the following "pseudo-game". From x_i^*, there does not exist an approval committee capable of initiating a transition to *any* other state. Notice that a Markovian equilibrium exists for this pseudo-game (see the Supplementary Notes for the general existence proof) and that x_i^* is absorbing. Denote by σ^* the equilibrium strategies for the pseudo-game. Return now to the actual game and suppose that players use the strategies σ^*; suppose also from x_i^*, that player i always proposes x_i^* and rejects *any* other transition. For other players $j \neq i$, any proposal and response strategies may be specified. Denote this new strategy profile σ'. Notice that σ^* and σ' specify the same transitions for the pseudo-game and actual game and no player has a profitable deviation from x_i^*. Therefore, σ' constitutes an equilibrium of the actual game. This equilibrium has an efficient absorbing state, x_i^*.

Case 2: For all i and for all $x_i^* \in X_i^*$, $|S^i(x_i^*)| = 1$. A number of subcases emerge:

(a) $\pi(x_1^*) = \pi(x_2^*) = \pi(x_3^*) = \pi_0$ for some (x_1^*, x_2^*, x_3^*), but the maximizing structures are not necessarily unique.

(b) $\pi(x_1^*) = \pi(x_2^*) = \pi_0$ and $\pi(x_3^*) = \pi_3$, and while the maximizing structures are not necessarily unique, Case 2(a) does not apply.

(c) For all players $i = 1, 2, 3$, π_i is the unique maximizing structure.

(d) $\pi(x_1^*) = \pi_0$, $\pi(x_j^*) = \pi_j$, $j = 2, 3$ and each maximizing structure is unique.

We now prove the proposition for each of these cases.

Case (a). Here x_0, the unique state corresponding to π_0, is weakly Pareto-dominant and we construct an equilibrium as follows. From any state x, every player proposes a transition to x_0 and every player accepts this proposal. A deviant proposal y is accepted if $V_i(y) \geq V_i(x) = (1 - \delta)u_i(x) + \delta u_i(x_0)$. This is clearly an equilibrium with x_0 efficient and absorbing.

Case (b). The proof is similar to Case 1. Consider a pseudo-game in which there is no approval committee that can initiate a transition away from x_0. Again, we are assured of a Markovian equilibrium for the pseudo-game; denote the equilibrium strategies by σ^* and notice that x_0 is absorbing. In the actual game, suppose that all players use the strategies given by σ^*, and suppose that, at x_0, players 1 and 2 always propose x_0 and reject any transition from x_0. Call these strategies σ'.

As in Case 1, notice that σ^* and σ' specify the same transitions for the pseudo-game and the actual game, and no player has a profitable deviation from x_0. Therefore, σ' constitutes an equilibrium of the actual game.

The following preliminary result will be useful for cases (c) and (d):

LEMMA 10.1. *Suppose that player i's maximizing structure $\hat{\pi}$ is unique, and that $\hat{\pi} \in \{\pi_0, \pi_i\}$. Let $Y = \{y \mid \pi(y) \in \{\pi_0, \pi_i\} - \hat{\pi}\}$. Then in any equilibrium such that $x \in X_i^*$ is not absorbing, $V_i(x) > V_i(y)$ for all $y \in Y$.*

Proof. We prove the case for which $\hat{\pi} = \pi_i$. The proof of the case for which $\hat{\pi} = \pi_0$ is identical. Let $x \in X_i^*$. Note that $Y = \{x_0\}$. Suppose on the contrary that $V_i(x_0) \geq V_i(x)$. We know that

$$V_i(x) = (1 - \delta)\bar{u}_i + \delta \int_X V_i(z)P(x, dz).$$

Now, there could be — with probability μ — a transition to the singletons, which player i need not approve. All other transitions must be approved by i, and she must do weakly better after such transitions. Using $V_i(x_0) \geq V_i(x)$, it follows that

$$V_i(x) \geq (1 - \delta)\bar{u}_i + \delta [\mu V_i(x_0) + (1 - \mu)V_i(x)] \geq (1 - \delta)\bar{u}_i + \delta V_i(x),$$

so that $V_i(x) \geq \bar{u}_i$. Strict inequality is impossible since $\bar{u}_i$ is i's maximal payoff. So $V_i(x) = \bar{u}_i$, but this means that x is absorbing.   □

The next two lemmas prepare the ground for case (c).

LEMMA 10.2. *Assume Case 2(c). Let y be not absorbing, and $\pi(y) = \pi_j$. Then y transits one-step to x_0 with positive probability.*

Proof. Suppose not. Then player j is on the approval committee for every equilibrium transition from y. Therefore

$$V_j(y) \geq u_j(y).$$

At the same time, y is not absorbing by assumption. But then the above inequality is impossible, since $u_j(y)$ is the uniquely defined maximal payoff for j across all coalition structures. □

LEMMA 10.3. *Assume Case 2(c). Suppose that a state x^i, with coalition structure π_i, is part of a nondegenerate recurrence class (starting from x^i). Then $V_j(x_0) = V_j(x^i)$ for all $j \neq i$.*

Proof. First, since x^i is not absorbing, by Lemma 10.2, x^i transits one-step to x_0 (with positive probability) and both players $j \neq i$ must approve this transition. Therefore

(10.4) $$V_j(x_0) \geq V_j(x^i).$$

Next, consider a path that starts at x_0 and passes through x^i (there must be one because x^i is recurrent). Assume without loss of generality that it does not pass through x_0 again. If both individuals $j \neq i$ must approve every transition between x_0 and x^i, we see that $V_j(x_0) \leq V_j(x^i)$, and combining this with (10.4), the proof is complete.

Otherwise, some $k \neq i$ does not need to approve some transition. This can only be a transition from x_0 to a state x^k with coalition structure π_k, with subsequent movement to x^i without reentering x_0. So $V_k(x^i) \geq V_k(x^k)$. But x^k itself is not absorbing and so by Lemma 10.2 transits one-step to x_0 (with positive probability). By Lemma 10.1, $V_k(x^k) > V_k(x_0)$. Combining these two inequalities, $V_k(x^i) > V_k(x_0)$, but this contradicts (10.4). □

Case (c). We divide up the argument into two parts. In the first part, we assume that for some i, some state x^i (with coalition structure π_i) is part of a nondegenerate recurrence class. Suppose that no efficient payoff limit exists. We first claim that

(10.5) $$V_j(x_0) = V_j(x^i) = u_j(x^i) \text{ for all } j \neq i.$$

To prove this, consider an equilibrium path from x^i. If this path *never* passes through x_0, then it is easy to see that all three players must have their value functions monotonically improving throughout, so one-period payoffs converge. Moreover, the limit payoff for player i must be at the maximum, so this limit is efficient. Given our presumption that there is no efficient limit, the path does pass through x_0, so consider these alternatives:

(i) For some $j \neq i$, the path passes a state y^j (with structure π_j) before it hits x_0. Moreover, y^j is not absorbing, and so by Lemma 10.2 it must transit one-step to x_0 with positive probability. However, player i must approve all these moves; so $V_i(x_0) \geq V_i(y^j) \geq V_i(x^i)$. But this contradicts Lemma 10.1. So this alternative is ruled out.

(ii) Otherwise, the path *either* transits one-step to x_0, *or* passes through a sequence of moves, *all* of which must be approved by both players $j \neq i$. So for any one-step transition from x^i to a state y, we have

$$V_j(x_0) \geq V_j(y) \geq V_j(x^i)$$

for $j \neq i$. But by Lemma 10.3, $V_j(x_0)$ *equals* $V_j(x^i)$ for $j \neq i$. It follows that for every one-step transit y,

$$V_j(y) = V_j(x^i)$$

for $j \neq i$. Consequently, for each such j,

$$V_j(x^i) = (1 - \delta)u_j(x^i) + \delta \int V_j(y)P(x^i, dy) = (1 - \delta)u_j(x^i) + \delta V_j(x^i).$$

Using this, and $V_j(x_0) = V_j(x^i)$ for $j \neq i$, the claim is proved.

We now show that there is an efficient absorbing state, contrary to our initial presumption. Consider a state x^i (with structure π_i) to which the claim just established applies. By condition T, there is some other state x^*, also with coalition structure π_i, such that for some $j \neq i$, $u_j(x^*) > u_j(x^i)$. Because j must approve every transition from x^*, $V_j(x^*) \geq u_j(x^*) > u_j(x^i) = V_j(x_0)$, where this last equality uses the claim. So x^* cannot have an equilibrium transition to x_0, but then i must approve *every* equilibrium transition. However, since π_i gives player i his unique maximal payoff, he will reject every transition to a different coalition structure. Therefore, x^* is both absorbing and efficient.

For the second part of case (c), suppose now that all recurrence classes are singletons. Assume by way of contradiction that all these are inefficient. This immediately rules out all absorbing states x^i with $\pi(x^i) = \pi_i$ for some i, and it also rules out x_0.[11]

Now consider any state x^i with $\pi(x^i) = \pi_i$. Since it is not absorbing, $V_j(x^i) \geq u_j(x^i)$ for $j \neq i$. Also Lemma 10.2 tells us that x^i transits one-step to x_0 with positive probability, so $V_j(x_0) \geq V_j(x^i) \geq u_j(x^i)$. In particular, $V_j(x_0) \geq \max\{\bar{u}(j, \pi_i), i \neq j\}$ for all $j = 1, 2, 3$. Moreover, because π_j is maximal for j and j must approve all other transitions from x_0 (as well as to all states from the structure π_j except for x_0), we have $V_j(x_0) \geq u_j(x_0)$, so $V_j(x_0) \geq \max\{u_j(x_0), \bar{u}(j, \pi_i), i \neq j\}$ for all $j = 1, 2, 3$. Since x_0 and x^i are transient, there must be a path from x_0 to an absorbing state x_G, but this implies that any

---

[11]If x_0 is absorbing and inefficient, then it is dominated either by a state for the grand coalition or by a state with coalition structure π_i for some i. Either way, by minimality of approval committees, x_0 will surely fail to be absorbing.

such absorbing state must satisfy $u_j(x_G) \geq \max\{u_j(x_0), \overline{u}(j, \pi_i), i \neq j\}$ for all $j = 1, 2, 3$. Therefore, x_G is efficient, contradicting our initial supposition.

Case (d). Proceed again by way of contradiction; assume there is no Markov equilibrium with efficient absorbing payoff limit. It is immediate, then, that any state x such that $\pi(x) \in \{\pi_0, \pi_2, \pi_3\}$ is not absorbing. It also gives us the following preliminary result:

LEMMA 10.4. *If any state x^1 with $\pi(x^1) = \pi_1$ is absorbing, then x^1 is not dominated by any state y with $\pi(y) \in \{\pi_0, \pi_G\}$.*

Proof. Suppose this is false for some x^1. It is trivial that x^1 cannot be dominated by any grand coalition state; otherwise x^1 wouldn't be absorbing. So x_0 dominates x^1. Consider any player $j \neq 1$. From x_0, there may be a transition to z^j with $\pi(z^j) = \pi_j$, which j need not approve. She must approve *all* other transitions from x_0. Thus, along the lines of Lemma 10.1, we see that $V_j(x_0) \geq u_j(x_0) > u_j(x^1) = V_j(x^1)$ for $j = 2, 3$, but this contradicts the presumption that x^1 is absorbing (given the minimality of approval committees, 2 and 3 will jointly deviate).   □

As in case (c), divide the analysis into different parts.

(i) All recurrence classes are singletons.

Since all absorbing states are assumed inefficient, it is clear that all absorbing states must either have coalition structure π_1 or π_G (since all states with coalition structures π_0, π_2 and π_3 are efficient). Consider x_0; it is transient. Let $\hat{x}$ be an absorbing state reached from x_0. By Lemma 10.2, we know that there must be a transition from any state with coalition structure π_2 or π_3 to x_0, and — because π_0 is maximal for player 1 — from x_0 to some state with coalition structure π_1 with strictly positive probability. Therefore, we may conclude that

$$
\begin{aligned}
V_1(x_0) &\geq \max\{\overline{u}(1, \pi_2), \overline{u}(1, \pi_3)\} \\
(10.6) \quad V_2(\hat{x}) &\geq V_2(x_0) \geq \max\{\overline{u}(2, \pi_3), u_2(x_0)\} \\
V_3(\hat{x}) &\geq V_3(x_0) \geq \max\{\overline{u}(3, \pi_2), u_3(x_0)\}.
\end{aligned}
$$

This implies that $\hat{x}$ is not Pareto-dominated by π_0, π_2 or π_3.

Now, if $\pi(\hat{x}) = \pi_1$, then Lemma 10.4, the fact that π_0, π_2 and π_3 are not absorbing, and (10.6) allow us to conclude that $\hat{x}$ must be efficient, a contradiction. So suppose that $\pi(\hat{x}) = \pi_G$. Note that $\hat{x}$ cannot be dominated by a state y such that $\pi(y) = \pi_1$. For $V_i(y) \geq u_i(y)$ for $i = 2, 3$. Moreover, an argument along the same lines as Lemma 10.1 easily shows that $V_1(y) \geq u_1(y)$. Therefore, if $\hat{x}$ were dominated by y, there would be a profitable move from $\hat{x}$, contradicting the presumption that it is absorbing. Therefore,

$\hat{x}$ must be efficient, but this contradicts our assumption that no absorbing state is efficient.

(ii) There is some nondegenerate recurrence class (and all other states are either transient or inefficient).

Observe that analogues to Lemmas 10.2 and 10.3, and the first part of case 2(c) can be established for case (d). However, whereas in case 2(c), we were able to pin down the equilibrium payoff of *two* players along some nondegenerate recurrence class, now we can only pin down the equilibrium payoff of *one* player; that is, for a recurrent class which transits from x_0 to x_i, $i \neq 1$, we have: $V_j(x_0) = V_j(x_i)$ and $V_j(x_0) = u_j(x_i)$ for $j \neq 1, i$.

Observe that if, for the player j whose payoffs we have pinned down, $u_j(x_i)$ *equals* $\bar{u}(j, \pi_i)$ *and* for the other player k who is part of the doubleton coalition with j, $V_i(x_0) \geq \bar{u}(k, \pi_i)$, the argument based on condition [T] will not go through.[12] That is, we cannot find another efficient state which one player (whose consent would be required for any transition) prefers to x_0. In this case, we must construct an equilibrium with some efficient absorbing state, and this is our remaining task.

First suppose that there does not exist a state x such that $u_i(x) > u_i(x_0)$ for $i = 2, 3$ and for all y such that $\pi(y) = \pi_1$, $u_1(x) \geq u_1(y)$.[13] In the construction of the equilibrium, the following sets of states will be important: $\{x_0\}$ and

$$S_2^u = \{x | \pi(x) = \pi_2, u_3(x) > u_3(x_0)\},$$
$$S_2^d = \{x | \pi(x) = \pi_2, u_3(x_0) \geq u_3(x)\},$$
$$S_3^u = \{x | \pi(x) = \pi_3, u_2(x) > u_2(x_0)\},$$
$$S_3^d = \{x | \pi(x) = \pi_3, u_2(x_0) \geq u_2(x)\},$$
$$S^u = \{x | \pi(x) \in \{\pi_1, \pi_G\}, x \text{ efficient}\},$$
$$S_1^D = \{x | \pi(x) \in \{\pi_1, \pi_G\}, \text{ there is } z \in \{S_2^u, S_3^u, S^u, x_0\} \text{ with } \mathbf{u}(z) \gg \mathbf{u}(x)\},$$
$$S_2^D = \{x | \pi(x) \in \{\pi_1, \pi_G\}, \text{ there is } z \in \{S_2^d, S_3^d\} \text{ with } \mathbf{u}(z) > \mathbf{u}(x); \text{ and there is } j \in \{23\} \text{ such that } u_j(x_0) < u_j(x)\} - S_1^D$$

Consider the following description of strategies:

(a) For all players $i = 1, 2, 3$, from x_0 all players offer x_0 and accept a transition to another state y only if $V_i(y) > V_i(x_0)$.

---

[12]Of course, if these conditions are not satisfied, the same argument as in 2(c) implies the existence of an efficient absorbing state.

[13]That is, there is no state that players 2 and 3 prefer to the singletons which they can achieve, either directly or indirectly (by initiating a preliminary transition to the coalition structure π_1).

(b) From all states $x \in S_2^d \cup S_3^d$, players i such that $|S^i(x) = 2|$ propose and accept x_0, while player i such that $|S^i(x)| = 1$ proposes the status quo. An arbitrary player k accepts a transition to another state y only if $V_k(y) > V_k(x)$.

(c) From all states $x \in S_2^u \cup S_3^u \cup S^u$, all players propose the status quo and an arbitrary player k accepts a transition to another state y only if $V_k(y) > V_k(x)$.

(d) From all states $x \in S_1^D$, all players propose a state $z(x) \in S_2^u \cup S_3^u \cup S^u \cup \{x_0\}$ and an arbitrary player k accepts a transition to another state y only if $V_k(y) > V_k(x)$. If x is dominated by x_0, we require $z(x) = x_0$.

(e) From all states $x \in S_2^D$, all players propose the status quo and an arbitrary player k accepts a transition to another state y only if $V_k(y) > V_k(x)$.

It is easy to see that these strategies constitute an equilibrium in which the singletons are absorbing. Moreover, every other state is either absorbing itself or transits (one-step with positive probability) to some absorbing state. Note that the states in S_2^D are absorbing for δ high enough and are inefficient. The reason they are absorbing is clear: if a transition to a dominating state were allowed, there would eventually be a transition to the singletons, which, by assumption, hurts one of the players whose original consent is needed. That the actions defined in (e) above constitute best responses for δ high enough follows from arguments similar to van Damme, Selten and Winter (1990): with a finite number of states and sufficiently patient players any such absorbing state could be implemented; one player will always prefer to reject any other offer.

Now suppose that there exists a state x such that $u_i(x) > u_i(x_0)$ for $i = 2, 3$ and for some y such that $\pi(y) = \pi_1$, $u_1(x) \geq u_1(y)$. In this case, the singletons clearly cannot be absorbing for δ high enough. However, with a finite number of states one can easily construct an equilibrium with a positive probability path from x_0 to some efficient absorbing state for δ high enough. In particular, from x_0, there is a positive probability transition to a state $y \in \pi_1$; from y there is a probability 1 transition to some efficient state y' which dominates y (if such a state exists; if not, y is absorbing).[14]

□

Appendix C

Proof of Observation 10.2.

---

[14]From x_0, there may also be a positive probability transition to some other state z. However, if $\pi(z) \in \{\pi_2, \pi_3\}$ it is clearly efficient since at these states players 2 and 3 obtain their unique maximum. Moreover, for δ high enough, it cannot be that $\pi(z) = \pi_G$, since then this would imply that z Pareto-dominates y'.

Step 1: x_3 and x_4 are not absorbing.

It is easy to see that $V_i(x_4) \geq 2$ for $i = 1, 2$. Moreover, since players 1 and 2 can initiate a transition from x_3 to x_4, x_3 is easily seen to be not absorbing. Similarly, $V_j(x_1) \geq 4$ for $j = 3, 4$; therefore, since players 3 and 4 can achieve x_1 from x_4, x_1 is not absorbing.

Step 2: x_2 absorbing implies x_2 is globally absorbing.

Suppose that x_2 is absorbing. Then clearly from x_1, players 1 and 2 would induce x_2. Moreover, since x_3 and x_4 are not absorbing, if x_2 is not reached, then x_1 must be reached infinitely often. But then 1 or 2 would get a chance to propose with probability 1 and would then take the state to x_2, a contradiction.

Step 3: x_2 cannot be globally absorbing.

If x_2 is globally absorbing then, from x_2, players 3 and 4 can get a payoff of 10 for some period of time, by initiating a transition to x_3, followed by, *at worst*, 2 for one period and 4 for another period, before returning to x_2, where it will get 5 forever thereafter.[15] This sequence of events is clearly better for players 3 and 4 than remaining at x_2.

Step 4: x_1 absorbing implies x_1 globally absorbing.

Steps 2 and 3 imply that x_2 cannot be absorbing. Moreover, Step 1 tells us that neither x_3 nor x_4 can be absorbing. In particular, from x_2 players 3 and 4 initiate a transition to x_3, from x_3 players 1 and 2 initiate a transition to x_4 and (at least) players 3 and 4 initiate a transition back to x_1. Therefore, if x_1 is absorbing, it is globally absorbing.

Step 5: Every equilibrium is inefficient.

First suppose that we had an equilibrium in which x_1 is not absorbing. Then from the above analysis, nothing is absorbing. Now consider x_2. If players 1 and 2 always accept an offer of a transition from x_1 to x_2, then 3 and 4 will strictly prefer to initiate a transition from x_2 to x_3: in so doing, they can achieve an average payoff of at least $\frac{10+2+4}{3} = \frac{16}{3} > 5$. However, it is easily seen that players 1 and 2 earn an average payoff strictly less than 4 in this case. Therefore, players 1 and 2 would rather keep the state at x_1, contradicting the presumption that x_1 was not absorbing.

---

[15]Surely, players 1 and 2 must initiate a transition to x_4 with some positive probability; otherwise, x_3 would be absorbing (which Step 1 shows to be impossible). However, once at x_4, under the assumption that any player can propose to move to any state, and the fact that (by Step 2) from x_1 there would be an immediate transition to x_2, there is no need to even pass through the intermediate state x_1.

The only remaining possibility is one in which players 1 and 2 are indifferent between a x_1 and x_2 and players 3 and 4 are indifferent between x_2 and x_3. If such an equilibrium were to exist, it must be that $V_i(x_1) = V_i(x_2) = 4$ for $i = 1, 2$, and $V_j(x_2) = V_j(x_3) = 5$ for $j = 3, 4$. Therefore, if such an equilibrium were to exist, it would also be inefficient since players spend a non-negligible amount of time at the inefficient states x_1 and x_4.

Thus either x_1 is the unique absorbing state or there is a sequence of inefficient cyclical equilibria depending on $\delta_n \nearrow 1$ such that players 1 and 2 are indifferent between x_1 and x_2 and players 3 and 4 are indifferent between x_2 and x_3.   □

Part 3

A Blocking Approach to Coalition Formation

CHAPTER 11

Blocking

This part of the book describes a second approach to coalition formation, one firmly grounded in the traditional theory of cooperative games. In what follows, there is no protocol — no explicit protocol anyway — that determines the choice of individual proposers and responders. In short, there is no game form in the sense that is commonly understood in noncooperative game theory. Rather, we join traditional cooperative game theory in moving one step away from methodological individualism, and regard *coalitions* as fundamental behavioral units. That is, coalitions form, agree, object or counterobject without a clear formal statement of which particular individual may be instigating such behavior. At the same time, we do not depart too far from an individual perspective: we presume that coalitions cannot act without the consent of their members.

11.1 The Core Revisited

Section 7.1 in Chapter 7 already introduces the concept of the core. Let's look at the concept more closely. Begin with a characteristic function $\mathbf{U}$ defined on a finite set N of n players: it specifies for every coalition S a set of payoff vectors $\mathbf{U}(S) \subseteq \mathbb{R}^S$ for that coalition. Recall that a characteristic function exhibits *transferable utility* (it is TU) if for each coalition S there is a number $v(S)$, describing the overall worth of that coalition, such that

$$\mathbf{U}(S) = \left\{ \mathbf{u} \in \mathbb{R}^S \mid \sum_{i \in S} u_i \leq v(S) \right\}.$$

More generally, $\mathbf{U}(S)$ is simply a set of payoff vectors $\mathbf{u} \in \mathbb{R}^S$ that are feasible for the coalition S. A payoff vector (or allocation) $\mathbf{u}$ on N is *feasible for the coalition structure π* if for every $S \in \pi$, $\mathbf{u}_S$ is feasible for S.

Let $\mathbf{u}$ be an allocation (feasible or not) for a set of players S. Say that $(T, \mathbf{u}')$ *blocks* $\mathbf{u}$ if T is a subcoalition of S ($\emptyset \neq T \subset S$), $\mathbf{u}'$ is feasible for T, and $\mathbf{u}' \gg \mathbf{u}_T$. In this case we say that $\mathbf{u}$ is *blocked* and that T is a *blocking coalition*. If there is no such blocking coalition, say that $\mathbf{u}$ is *unblocked*.

Notice that a block requires *every* member of the "blocking coalition" to be strictly better off. One might demand a weaker notion of blocking, in which all individuals are weakly better off, and at least one individual is strictly better off. More often than not, this change makes no difference.[1]

The blocking notion swiftly yields the concept of the *core of a coalition*, a fundamental solution concept in cooperative game theory. It is simply the set of all feasible allocations for that coalition that are unblocked. When we say "the core" without qualification, we typically refer to the core of the grand coalition N. It is also possible to define the core of a coalition *structure π*: it is the set of all unblocked allocations that are feasible for the structure π.

We've already described what it means for a TU characteristic function to be superadditive. The extension to general characteristic functions is straightforward. Say that a coalition S is *superadditive* if for every partition π of S and every allocation $\mathbf{u}$ feasible for it, there is an allocation $\mathbf{u}'$ feasible for S such that $\mathbf{u}' \geq \mathbf{u}$. A *superadditive characteristic function* is one which only has superadditive coalitions.

It is, of course, well known that superadditivity is not sufficient for the core of a coalition to be nonempty. We've already seen in Example 7.1 and Observation 7.1 of Chapter 7 that more is required. The celebrated theorem of Bondareva and Shapley provides a necessary and sufficient condition for core nonemptiness in TU games, called balancedness. The natural extension of this concept to general games is sufficient (though not always necessary) for the

---

[1]When payoffs are transferable, the two definitions are equivalent. But there are situations in which payoffs are not transferable, or effectively become nontransferable; see, e.g., the discussion following Example 11.1 below. Then the two definitions of blocking could have very different implications, as they do indeed in that example.

nonemptiness of the core in those games; this is the theorem of Scarf (1971). A precise description of the balancedness condition is not really necessary for our purposes; suffice it to note that it is stronger than superadditivity.

11.2 Farsightedness in Blocking

There are two reasons I resurrect this perfectly classical discussion of cores, empty or otherwise. One of them has been discussed at some length in Chapter 7: when the core is empty, traditional theory is silent on what exactly happens. *This* coalition (with an empty core) is "unstable", but which coalitions *do* form? The core concept does not address this question. But there is a second aspect which acquires particular importance in the context of our discussion of blocking. Following a block, it may be possible (by further "blocks") to keep moving to an allocation entirely different from the original proposal or the allocation that blocked it to begin with. Of what relevance, then is the original block?

Here is a related way of looking at the issue. The core is the set of all allocations which are unblocked by any subcoalition. *But this definition does not put subcoalitions to the test in the same way as it does the grand coalition.* Put yet another way, the definition of the core presumes that agents are not farsighted and do not see forward to the "ultimate" consequences of their own actions.

In the bargaining approach to coalition formation pursued in Part 2 of this book, this sort of farsightedness is very naturally built in. A proposal is made. That proposal may be rejected. But the rejector's actions are invariably followed by a fresh proposal, and the rejector knows this. The act of rejection is therefore "treated in the same way" as the events leading up to the original proposal. In a model based on blocking, however, such features will have to be introduced in a more explicit way.

11.3 A First Pass at Farsightedness

Suppose that we have a characteristic function, given by $\mathbf{U}$. Think of a solution concept F as a mapping that assigns a subset of allocations (possibly empty) to every coalition S: thus $F(S) \subseteq \mathbf{U}(S)$ for every S. For instance, in the case of the core, $F(S) = \text{Core}\,(S)$ is the set of all unblocked allocations. Formally:

Core $(S) = \{\mathbf{u} \in \mathbf{U}(S)|$ for no $T \subset S$ and $\mathbf{u}' \in \mathbf{U}(T)$ is $\mathbf{u}' \gg \mathbf{u}_T\}$.

Note that there are no constraints at all imposed on the potential blocking coalitions. It is sufficient for coalition T to drum up any allocation $\mathbf{z} \in \mathbf{U}(T)$. We might demand, however, that such an allocation in turn exhibit the same level of credibility that we demand of the "original" coalition S. There is an obvious circular feel to such a definition: the "credibility" of S is being tested, but only by similarly "credible" objections from T. Here is a more formal statement:

Core$^*(S) = \{\mathbf{u} \in \mathbf{U}(S)|$ for no $T \subset S$ and $\mathbf{u}' \in$ Core$^*(T)$ is $\mathbf{u}' \gg \mathbf{u}_T\}$.

Notice that "Core*" appears both on the left and right hand sides of the description. This is the implicit circularity I was discussing above. With only subsets doing the blocking, however, there is no fear of a conceptual impasse; one can simply define the concept recursively upwards, starting from singleton coalitions.

In principle, then, blocking is harder, so that the "credible core" defined in this way should be a superset of the core. But it turns out that the imposition of credibility in this way does nothing to change the core. The following observation (see Ray (1989)) is very simple:

OBSERVATION 11.1. Core$^*(S) =$ Core (S) *for all coalitions S.*

Proof. Clearly Core $(S) \subseteq$ Core$^*(S)$ for all S. Suppose that for some S, equality does not hold. Then there is $\mathbf{u} \in$ Core$^*(S)$−Core (S). Because $\mathbf{u} \notin$ Core (S), there is $T \subseteq S$ which blocks $\mathbf{u}$ using $\mathbf{u}' \in \mathbf{U}(T)$. Because $\mathbf{u} \in$ Core$^*(S)$, $\mathbf{u}' \notin$ Core$^*(T)$. So $\mathbf{u}'$ is blocked by some coalition W using an allocation $\mathbf{u}''$ in Core$^*(W)$. Now $W \subseteq S$. Moreover, it is easy to check that if $(W, \mathbf{u}'')$ blocks $(T, \mathbf{u}')$ while $(T, \mathbf{u}')$ blocks $(S, \mathbf{u})$, then $(W, \mathbf{u}'')$ blocks $(S, \mathbf{u})$. But this contradicts our presumption that $\mathbf{u} \in$ Core$^*(S)$, and the proof is complete. □

One piece of good news, then, is that the core has some credibility features already built in. If we alter the definition to permit only those blocks that are themselves immune to further blocking, the set of unblocked allocations should, in principle, expand. But it doesn't, and the reason it doesn't is that for every "noncredible" block, there is a further "credible" block that opposes not just the first block, but the original proposal itself. The solution *concept* changes, but the solution doesn't.

At the same time, don't be entirely taken in by the simplicity of the argument: even the simplest variants of it create entirely new outcomes. For instance, suppose that after a coalition discovers the set of its unblocked allocations, it implements only a particular subset of them. Formally, suppose that for every set of payoffs A (to be interpreted in the sequel as the "unblocked" set) for coalition S, there is a choice rule θ_S that picks out a subset of those outcomes: $\theta_S A \subseteq A$.

Why might a coalition do this? The answer could have to do with bargaining or norms. The coalition may internally bargain over the choice from the various unblocked allocations. If this is done and properly foreseen by all parties, the bargain must be taken into account when forecasting a coalitional block. Or the coalition may possess norms, having to do with egalitarianism or fairness, that forces it to select further from the options open to it.

Now define a credible-core-like solution concept relative to this choice rule; call it Σ. It has the property that for each coalition S, $\Sigma(S) = \theta_S A$, where

$$A = \{\mathbf{u} \in \mathbf{U}(S) | \text{For no } T \subset S \text{ and } \mathbf{u}' \in \Sigma(T) \text{ is } \mathbf{u}' \gg \mathbf{u}_T\}.$$

Variants such as these will generally have a significant impact on the solution. To see this, focus on a particular specification of θ_S: the *egalitarian solution*, studied by Dutta and Ray (1989, 1991).[2] Suppose that the underlying game is TU, and that for every coalition S and potential set of payoffs A, θ_S picks out the "most equal" allocations in A. Specifically, θ_S selects the Lorenz-maximal elements of A.

EXAMPLE 11.1. *There is a set N of players, which can be split into two groups as follows: $N = L \cup R$, where $|R| = |L| - 1 > 0$. For any coalition S, $v(S) = \min\{|S \cap L|, |S \cap R|\}$.*

This example is a classical one in cooperative game theory, and it is known as the "right and left gloves" situation. Each player owns one unit each of *one* of two kinds of inputs (a left-handed glove or a right-handed glove), and one unit of a consumable good can be produced only by joining together one unit each of the two inputs.

It is well known that the core of this game contains a single payoff allocation. That allocation gives one unit each to members of R,

---

[2]We've already encountered the egalitarian solution in the context of coalitional bargaining; see Section 7.4.4 in Chapter 7.

and none to members of L, even if L and R are both very large sets (so that relatively speaking, they are almost the same size). Indeed, this disconcerting example is sometimes put forward as an implicit criticism of the core concept.

However, suppose that every coalition is restricted to be egalitarian, in the sense that ex-post, they invariably try to attain as equal a division of their worth consistent with the no-blocking requirement. Then we may use the solution concept Σ relative to θ_S being the egalitarian rule, as described above. Proceed recursively, starting from two-person coalitions. They must always divide their worth equally. Now consider a three-person coalition, with two Ls in them and just one R. The worth of this coalition is 1. What is the set of its unblocked allocations? It is precisely the set

$$\left\{ \mathbf{u} \in \mathbb{R}^3 \mid \sum_i u_i = 1; \text{ for no } i \in L \text{ and } j \in R \text{ is } (1/2, 1/2) \gg (u_i, u_j) \right\}.$$

We must apply θ to this set in order to obtain its Lorenz-maximal element. There is a unique solution: give the R-type half the worth, and the L-types a quarter each. Now we can build up to larger coalitions. For each coalition, define a type (L or R) to be a *minority* if it is not more than half the population in the coalition, and a majority otherwise. We then have, via a similar process of computation,

OBSERVATION 11.2. *For every coalition S in Example 11.1, there is a unique allocation in $\Sigma(S)$, given as follows: $1/2$ for each minority type and $m/2(s - m)$ for each majority type, where s is the cardinality of S and m the size of the minority.*

Applying this observation to the grand coalition, we see that for large numbers of players, the result is almost equal division. The Rs, who are in the minority, obtain $1/2$ each, while the Ls, who are in the majority, obtain close to $1/2$ each: $|L|/2(|L| + 1)$, to be precise. Unlike the core, which allocates payoffs in dramatically uneven fashion when group composition is only slightly asymmetric, our solution changes very little from fully equal treatment.

It isn't my goal, however, to dwell on the philosophical merits of this solution *vis-a-vis* the core. The point is simply to note that the imposition of farsighted consistency can change the solution a lot, sometimes even picking out — as in this case — noncore allocations. Such allocations cannot be blocked, because we have imposed (consistent) restrictions on the credibility of the blocking process.

This will be one important feature of the theory we develop in the next chapters, but there is more to consider.

11.4 Externalities and Farsightedness

By far the more important consideration emerges once we consider externalities across coalitions. Now a characteristic function is no longer enough, and we must account for intercoalitional effects on payoffs. As we've already seen in Section 2.4, this raises a fundamental prediction problem.

When there are no externalities, a blocking coalition must indeed worry about which blocks are "credible". That is what we've just discussed. With externalities, the group must *also* attempt to predict the coalition structure that arises elsewhere. With no such prediction in hand, it will simply be unable to compute its own worth, let alone begin to think about which particular allocations are credible in the sense discussed in this chapter.

As we have discussed already, this is a conceptually distinctive problem which requires not just the presence of externalities but the fact that negotiations are underway to implement an allocation. If those negotiations break down, it is common knowledge that they have done so, and a deviating coalition must presume that the complementary part of the group will suitably rearrange themselves. In contrast, in theories of simultaneous, nonbinding play, these deviations do not have to "anticipate" that the remaining players will rearrange themselves, or indeed even change their actions! There is nothing to "anticipate".

This is the starting point for the methodology that we develop in the rest of the book. It isn't that we haven't encountered this issue before; of course we have. But so far we have concentrated on explicitly defined noncooperative games of negotiation. The project at hand is to extend the tools of cooperative game theory, and so approach the same problem with a different set of techniques.

11.5 Summary

This chapter begins a study of coalition formation by extending the techniques of cooperative game theory. A central notion is one of blocking, and the attendant concepts that stem from it — blocking coalitions and the core — are important ideas in the theory of cooperative games.

Starting with the core, we introduce two central ingredients of our approach to coalition formation. Both have to do with farsightedness. First, we explore the idea that a deviating coalition should not simply look at the *immediate* consequences of its own actions. Further deviations may follow, and the coalition should look ahead to the "final" consequence of its original deviation.

Second, we recall the fundamental prediction problem. When there are externalities, a deviating coalition must also attempt to forecast what its complementary coalition might do, in an attempt to predict its own worth. These two features will be central to the two chapters that follow.

We proceed along lines parallel to the chapters based on bargaining in Part 2. First we study irreversible agreements, in which coalitions that once agree to form cannot continue to renegotiate. Then we consider the problem of coalition formation in real time, in which constant renegotiation is possible.

My objective in this part of this book is not to provide an exhaustive analysis of the cooperative game-theoretic model. I simply put the model up as a reasonable alternative to the extensive-form bargaining approach followed so far, and I indicate some of the salient points. The chapters that follow are largely based on Ray and Vohra (1997) and Konishi and Ray (2003), but the reader interested in various other aspects of this literature should also consult the references in these chapters, on which I often rely heavily.

CHAPTER 12

Irreversible Commitments

We return to the fundamental backdrop for all that we do, described in Chapter 3. We have, then, a game Γ in strategic form. N is a set of players. Player i has action set A_i. $\mathbf{A}$ denotes the product of all action sets. Player i has a payoff function u_i defined on A.

A *coalition* is just a nonempty subset of N. As always, interpret a coalition to mean a set of players who are willing signatories to a binding agreement. The restriction of the product set A to coalition S will be denoted by $\mathbf{A}_S$. Similarly, we denote $(u_i(\mathbf{a}))_{i \in S}$ by $u_S(\mathbf{a})$.

Finally, a partition π of N into coalitions will be called a *coalition structure*.

As in the preceding chapters, we approach the coalition formation problem in two stages. Each player must forecast the interactive consequences of every conceivable coalition structure. Coalition formation occurs "at an earlier stage" with this "second-stage consequence" firmly in mind. In Chapter 3, we describe the second stage as a "coalitional equilibrium" in the spirit of Nash. Suppose that π is a coalition structure. An action vector $\mathbf{a}$ is a *coalitional equilibrium* (relative to π) if for no coalition $S \in \pi$ is there an action vector $\mathbf{a}'_S \in \mathbf{A}_S$ with $u_S(\mathbf{a}'_S, \mathbf{a}_{-S}) \gg u_S(\mathbf{a})$.

For singleton coalition structures a coalitional equilibrium is just a Nash equilibrium, and for the grand coalition it is simply the Pareto frontier of the game.

Proposition 3.1 of Chapter 3 show that under general conditions, a coalitional equilibrium always exists. For each coalition structure π, denote by $\beta(\pi)$ the set of coalitional equilibrium action vectors. We are now in a position to describe the "first-stage" theory of coalition formation from a cooperative perspective.

12.1 Equilibrium Binding Agreements

In this section, I follow Ray and Vohra (1997) in defining the set of *equilibrium binding agreements* $\mathcal{B}(\pi)$ for each coalition structure π. A typical binding agreement is, of course, a full specification of "equilibrium actions" taken by each of the agents, one specification for each coalition structure. If equilibrium binding agreements do exist for a given coalition structure π, we shall refer to π as an *equilibrium coalition structure*.

A central feature of what follows is the possible formation of new coalition structures from old ones. In keeping with the spirit of the core concept, we only permit new coalitions to form via the disintegration of existing coalitions. In a later section, we discuss Diamantoudi and Xue (2007), who drop this restriction.

Consider two coalition structures π and π', with π' a refinement of π. Think of having "moved" from π to π' by the formation of one or more new coalitions, each a subset of some element of π. Some of these coalitions may be thought of as "active movers", or *perpetrators*, in the creation of π', and others might be residual coalitions, or simply *residuals*, of individuals left behind by the perpetrators. Observe that we cannot uniquely identify a class of perpetrators. But we *can* say this: if a coalition in π breaks into n new coalitions, $n - 1$ of them must be labeled perpetrators, and the remaining coalition must be taken to be a residual. A *collection of perpetrators and residuals in the move from π to π'* is any labeling of the relevant elements of π' which satisfies the requirement in the previous sentence.

Fix a collection of perpetrators and residuals in the move from π to π'. A *re-merging* of π' is a coalition structure $\hat{\pi}$ formed by merging any collection of perpetrators with their respective residuals. Below, this will be used to capture situations in which some perpetrators contemplate not moving to π'.

I now recursively define equilibrium binding agreements (EBA) — call these $\mathcal{B}(\pi)$ — for each structure π. Two initial conditions will be used to set the recursion going. First, begin with the finest possible coalition structure π^* of singleton coalitions. In this case, we simply set $\mathcal{B}(\pi^*) = \beta(\pi^*)$, the set of coalitional equilibrium actions under π^*. Notice, moreover, that these are just the set of Nash equilibrium action vectors of the underlying game.

Next, consider coalition structures π which have π^* as their only refinement. Let $\mathbf{a} \in \beta(\pi)$. Say that $(\pi^*, \mathbf{a}^*)$ *blocks* $(\pi, \mathbf{a})$ if $\mathbf{a}^* \in \mathcal{B}(\pi^*)$ and there exists a perpetrator S such that $u_S(\mathbf{a}^*) \gg u_S(\mathbf{a})$.

Now for the recursion. Suppose that for some π the set $\mathcal{B}(\pi')$ has already been defined for all π' that are refinements of π. Moreover, assume that for each such π' and each $\mathbf{a}' \in \beta(\pi')$ we have described all $(\pi'', \mathbf{a}'')$ that block $(\pi', \mathbf{a}')$.

Let $\mathbf{a} \in \beta(\pi)$. Say that $(\pi, \mathbf{a})$ is *blocked* by $(\pi', \mathbf{a}')$ if π' is a refinement of π, and there exists a collection of perpetrators and residuals in the move from π to π' such that

[E.1] $\mathbf{a}'$ is an EBA for π': $\mathbf{a}' \in \mathcal{B}(\pi')$.

[E.2] There is a *leading perpetrator* S which gains from the move: $u_S(\mathbf{a}') \gg u_S(\mathbf{a})$, and

[E.3] Any re-merging of the *other* perpetrators is blocked by $(\pi', \mathbf{a}')$ as well, with one of these perpetrators as a leading perpetrator. Formally, let $\mathcal{T}$ be the set of all perpetrators, other than S, in the move from π to π'. Let $\hat{\pi}$ be a coalition structure formed by merging some of the elements of $\mathcal{T}$ with their respective residuals. Then $\mathcal{B}(\hat{\pi}) = \emptyset$ and there is $\hat{\mathbf{a}} \in \beta(\hat{\pi})$ and $S' \in \mathcal{T}$, such that $(\hat{\pi}, \hat{\mathbf{a}})$ is blocked by $(\pi', \mathbf{a}')$ with S' as the leading perpetrator.

We may now complete the recursion. A strategy profile $\mathbf{a}$ is an *EBA for π* if $\mathbf{a} \in \beta(\pi)$ and there is no $(\pi', \mathbf{a}')$ that blocks $(\pi, \mathbf{a})$. Denote by $\mathcal{B}(\pi)$ the set of all EBAs for π.

Thus, objections or blocks are defined perfectly consistently. A perpetrator can only expect to induce an equilibrium binding agreement in some refinement of the going coalition structure π; this is analogous to the credible core studied in the previous chapter, except that we take cross-coalitional externalities on board as well. Moreover, if this refinement involves the defection of *other* subcoalitions, conditions must be imposed that make it worthwhile for such coalitions to have defected. [E.3] captures this. To see this, observe that a re-merging — while it excludes the original leading perpetrator by definition — partially reverses the defection process, returning to intermediate coalition structures of the form $\hat{\pi}$. What [E.3] states is that each such merger should lack the ability to write EBAs, and moreover that there is some coalitional equilibrium under $\hat{\pi}$ which is blocked by the original defection(s).

Typically, many coalition structures admit EBAs . Which of these should be considered as *the* set of EBAs for the game? The answer to this question depends on what we consider to be the "initial" coalition structure under which negotiations commence. In keeping with the spirit of our exercise, which is to understand the outcomes of free and unconstrained negotiation, we take it that the initial structure is the grand coalition itself. Under this supposition, it is natural to focus on the set of equilibrium binding agreements for the grand coalition, or, if this set is empty, on the next level of refinement for which the set of EBAs is nonempty.

12.2 Farsightedness and Prediction

Our definition of blocking attempts to incorporate, at one stroke, the two related features of farsightedness and prediction. The farsighted perpetrator must look ahead to the "ultimate" consequences of her deviation, accounting for additional deviations that may follow her own. The prediction problem requires, in addition, that the perpetrator attempt to forecast what the ambient coalition structure will be when the dust has died down. We discuss these two features of the definition in this section.

12.2.1 Farsightedness and Sequential Blocking. Condition E.3 in our definition requires that if one or more (nonleading) perpetrators from a blocking partition are reunited with their erstwhile compatriots, the resulting "remerged" coalition structure is also blocked by the original blocking allocation. The idea is that other perpetrators cannot help but go along with the leading perpetrator. This attempts to capture the essence of farsightedness.

More concretely, blocking may be viewed as a *sequence* of acts, in which one coalition's proposal may be "counterattacked" by another, perhaps over several rounds. Such sequentiality is the basic idea behind the bargaining set of Aumann and Maschler (1964) and the consistent bargaining set of Dutta, Ray, Sengupta and Vohra (1989). The problem with these papers is that they do a bad job of understanding farsightedness along a sequence: a blocking coalition looks only at the immediate consequence of its block. If the coalition knows that an entire chain is involved, it should look instead at the "ultimate" consequence of its behavior. This idea goes back at least to Harsanyi's (1974) criticism of the notion of stable sets, and is further developed in Chwe (1994). Chwe argues for a sequential notion of blocking in which all agents look ahead to the end result of their actions, and not the state immediately following their block, which may be entirely irrelevant.

Apply that idea here. Suppose that $\mathbf{a} \in \beta(\pi)$ is a coalitional equilibrium under π. $(\pi', \mathbf{a}')$ is said to *sequentially block* $(\pi, \mathbf{a})$ if there exists a sequence $\{(\pi^0, \mathbf{a}^0), (\pi^1, \mathbf{a}^1), \ldots, (\pi^m, \mathbf{a}^m)\}$ such that:

[S.1] $(\pi^0, \mathbf{a}^0) = (\pi, \mathbf{a})$, $(\pi^m, \mathbf{a}^m) = (\pi', \mathbf{a}')$ and for every $i = 1, \ldots, m$, there is a coalition S^i such that S^i is the only perpetrator in the move from $(\pi^{i-1}, \mathbf{a}^{i-1})$ to $(\pi^i, \mathbf{a}^i)$.

[S.2] $\mathbf{u}_{S^i}(\mathbf{a}') \gg \mathbf{u}_{S^i}(\mathbf{a}^{i-1})$ for all $i = 1, \ldots, m$.

This notion of blocking contains an explicitly sequential account of how coalitions move, unlike our definition. Nevertheless, our notion of blocking subsumes its sequential counterpart.

OBSERVATION 12.1. *If* $(\pi', \mathbf{a}')$ *blocks* $(\pi, \mathbf{a})$, *then* $(\pi', \mathbf{a}')$ *sequentially blocks* $(\pi, \mathbf{a})$.

The proof, which I relegate to an appendix, isn't hard at all. It shows how the blocking notion that I use compresses *chains* of reasoning into a single step, captured by [E.2] and [E.3]. If $(\pi', \mathbf{a}')$ blocks $(\pi, \mathbf{a})$, then one can use this fact to construct a "path" linking the latter to the former, consisting of a single perpetrator at each stage of this path, so that all the perpetrators prefer the end result $(\pi', \mathbf{a}')$ to their starting points.

Indeed, the converse of this observation is generally not true. The blocking notion I use encompasses more than a single chain, for it allows the perpetrators to "reorganize" themselves in a variety of ways, thus allowing for a diversity of possible paths (see [E.3]). Using this intuition, Diamantoudi and Xue (2007) show that it is possible to fully characterize the blocking notion used here by using sequential blocking along a multiplicity of paths. I omit the details here, but will return to sequential blocking in Section 12.5 below.

12.2.2 The Prediction Problem. Buried within the blocking definition is also a prediction for the equilibrium coalition structure once a leading perpetrator precipitates a deviation. If $(\pi', \mathbf{a}')$ blocks $(\pi, \mathbf{a})$, then π' is precisely that prediction, and it is so on two grounds. First, $\mathbf{a}'$ is an equilibrium binding agreement under the structure π', which means that when π' is in turn subjected to the same treatment currently given to π, it will not disintegrate further. Second, no "halfway house" between π and π' is a natural stopping point: by the condition E.3, that halfway house would continue to split up.

As I've explained earlier, cooperative game theory, based as it is on the externality-free characteristic function, has largely ignored such matters. And when the subject has come up, the solution has been *ad hoc*. A leading instance of this is the important paper by Hart and Kurz (1983), which represents an early and significant contribution to the theory of binding agreements starting directly from the strategic form. The objective is to obtain and characterize "stable coalition structures", just as ours is here, and they run (as they must) into the prediction problem. To address it, they consider two notions of coalitional stability. The first corresponds to a strong equilibrium of a game in which a deviation by a coalition $T \subset S$ leaves $S - T$ as a residual, and all other coalitions remain unchanged. Under the second, when a coalition $T \subset S$ deviates, the members of $S - T$ break up into singletons, while all other coalitions remain the same.[1] Clearly, neither of these formulations satisfactorily addresses the prediction problem: there is no reason to believe either that the residual left by a deviating coalition will stay in one piece, or that it will break up into individual players.

The landmark paper which does attempt to provide a fully specified solution to the prediction problem is Aumann and Myerson (1988). But their approach is not within the blocking-based paradigm of traditional cooperative games. They study an extensive-form game in which where a given rule of order specifies the sequence in which players are allowed to form links. They study the noncooperative equilibrium of such a game. Because of their particular methodological emphasis, this paper is connected more closely to the bargaining

---

[1]They consider other equilibrium notions as well, based on the α-core and the β-core of the strategic game. For other examples in which the prediction problem needs to be addressed, see Dutta and Suzumura (1993), Chander and Tulkens (1995) and Carraro and Siniscalco (1993).

models studied earlier in Part 2 rather than cooperative game theory of the classical variety.

12.3 Inefficiency

Our definition permits the writing of *any* agreement to which players can jointly agree. Yet, a central theme that runs through the book — and this chapter is no exception — is that despite the ability to write agreements, inefficient outcomes are possible. As in Part 2, it is possible to write down several economic examples in which this happens. But my simple objective in this section is to record the fact that such situations are robust to arbitrary small perturbations of the underlying game, unrestricted in any way by the underlying economic context.

To be sure, we are unwilling to seriously consider instances of inefficiency that rely on incompatibly optimistic views of blocking. Such examples rely on the multiplicity of equilibria following a block, each coalition optimistically anticipating the equilibrium most beneficial to it. Consider, for example, a modified version of the battle of the sexes, with two players and two pure Nash equilibria yielding payoff vectors of, say, $(5, 1)$ and $(1, 5)$. Suppose, now, that there are (non-Nash) payoffs that Pareto-dominate either of these two outcomes, but do not dominate the vector $(5, 5)$. This game has no efficient binding agreement. However, such outcomes are obviously not robust to reasonable alternative definitions that rely on a lesser degree of optimism.[2]

We therefore demand of inefficient outcomes that they be compatible with the "essential uniqueness" of coalitional equilibrium for every coalition structure. Loosely, we ask that equilibrium payoffs be unique modulo any (externality-free) transfers of payoffs among players within any coalition. For a more precise description, consult Section 3.2.4.

---

[2]Variants of our definition can easily incorporate increasing degrees of pessimism, culminating in the requirement that a leading perpetrator must be better off in *every* equilibrium binding agreement of *every* coalition structure induced by it. As explained in Ray and Vohra (1997), such variants have their own potential drawbacks. In the end, we feel that none of this poses a serious problem provided that we apply the concept in a way that's sensitive to the potential pitfalls induced by multiplicity. Recall that the same issue came up in our discussion of the partition function; see Section 3.2.4.

Fix a finite action set for each player. Then the set of games may be identified with an appropriate Euclidean space of payoff profiles. In the proposition below, we refer to open subsets of games under this identification.

PROPOSITION 12.1. *Suppose that there are at least three players with at least three actions each. Then there exists an open set of games such that every game in this set satisfies essential uniqueness, and no EBA in any game is efficient.*

Observe that with essential uniqueness, every EBA for the grand coalition must necessarily be efficient in a two-player game. This justifies the use of at least three players in the proposition.[3]

To prove this proposition, consider the case of exactly three players with three actions each; the remaining cases are handled by a simple extension. In the following strategic form, player 1 chooses rows, 2 chooses columns and 3 matrices.

		x_{2a}	x_{2b}	x_{2c}
x_{3a}	x_{1a}	2.6,2.6, 2.6	3.2, 2.2, 3.2	3.7, 1.7, 3.7
	x_{1b}	2.2, 3.2, 3.2	2.7, 2.7, 3.7	3.1, 2.1, 4.1
	x_{1c}	1.7, 3.7, 3.7	2.1, 3.1, 4.1	2.6, 2.6, 4.6

		x_{2a}	x_{2b}	x_{2c}
x_{3b}	x_{1a}	3.2, 3.2, 2.2	3.7, 2.7, 2.7	4.1, 2.1, 3.1
	x_{1b}	2.7, 3.7, 2.7	3.1, 3.1, 3.1	3.6, 2.6, 3.6
	x_{1c}	2.1, 4.1, 3.1	2.6, 3.6, 3.6	2.9, 2.9, 3.9

		x_{2a}	x_{2b}	x_{2c}
x_{3c}	x_{1a}	3.7, 3.7, 1.7	4.1, 3.1, 2.1	4.6, 2.6, 2.6
	x_{1b}	3.1, 4.1, 2.1	3.6, 3.6, 2.6	3.9, 2.9, 2.9
	x_{1c}	2.6, 4.6, 2.6	2.9, 3.9, 2.9	3.3, 3.3, 3.3

Every player i has a dominant strategy, x_{ia}. Thus the unique Nash equilibrium, and the only EBA for the singleton structure π^*, is (x_{1a}, x_{2a}, x_{3a}), which is Pareto-dominated by (x_{1c}, x_{2c}, x_{3c}).

Next, consider an intermediate structure; say $\pi = (\{1\}, \{23\})$ (the game is symmetric so there is no loss of generality in this choice). Because x_{1a} is dominant for 1, any $z \in \beta(\pi)$ must have $z_1 = x_{1a}$. Thus we need only look at the first row of each matrix. With this in mind,

---

[3]The proposition also assumes that there are at least three strategies for at least three of the players. This assumption cannot be dropped free of charge, but I do not know the extent to which it can be weakened.

it is easy enough to see that (x_{1a}, x_{2b}, x_{3b}) is a coalitional equilibrium. Moreover, this action profile cannot be blocked by a deviation to π^*. It is, therefore, an EBA.

Indeed, this is the only EBA for this coalition structure: in all other coalitional equilibria, at least one of players 2 and 3 receives less than 2.6, the unique Nash payoff, and so these equilibria will be blocked. By symmetry, then, every EBA for every intermediate structure is inefficient.

Now consider the grand coalition. For any strategy profile it must be the case that there exists a player, i, who gets less than 3.7. This player can then block by deviating to $(\{i\}, \{j, k\})$ and earning 3.7. In fact, this is the *only* coalition structure that i can induce by deviating from the grand coalition. Thus, the grand coalition must break up into some intermediate coalition structure with an inefficient equilibrium.

Since all coalitional equilibria are strict, an open set of payoff functions that yield the same qualitative outcomes can easily be constructed.

Finally, this example is easily extended to consider additional strategies and/or additional players; the details are left to a footnote.[4]

Notice that the three-person example used in the proof does not allow for transfers of payoffs within a coalition. The conclusion remains unchanged even if such transfers are permitted. The nontransferability of utility isn't needed for the breakdown of efficiency.

The reader familiar with the public goods application in Section 6.2 of Chapter 6 will immediately see the connection with this example. We have here a three-person structure where a player, by inducing the coalition structure consisting of just herself in one coalition, and the other two players in another, can do better than the average payoff to a player in the (efficient) grand coalition. The consequent inefficiency hinges on the fact that in such a case, the

---

[4]If any player has additional actions, simply set payoffs to *all* players equal to 0 when any of those actions are played. This specification guarantees a robust set of games satisfying the properties of the proposition. If there are additional players, include them as dummies as far as the first three players are concerned, and for each such dummy player, define his payoff to be 1 provided (x_{1c}, x_{2c}, x_{3c}) is played by the first three players, and zero otherwise. It is easy to see that this modification does not change the earlier conclusions.

other two players are better off staying together than also breaking apart. Observe that while the payoffs to the three potential singleton deviants jointly dominate the outcome that can be achieved by the grand coalition, the game is still superadditive in the sense that the grand coalition can still Pareto-dominate each inefficient outcome.

12.4 An Application to Political Party Formation

The notion of equilibrium binding agreements is one with widespread applicability. Unlike the core, which applies only to characteristic functions, this solution concept may be used to study situations in which there are externalities across coalitions: the domain of applicability is significantly broader.

In what follows, I apply the idea to a model of political party formation. The analysis draws heavily on Levy (2004).

Levy views political parties as coalitions of politicians; these will be our set of agents. When a collection of parties compete for election, they play a "platform game", in which each party chooses whether or not to participate and (conditional on running) a platform or policy. The platform with the largest votes wins and is implemented. This game corresponds to our stage game, and its equilibria correspond to our coalitional equilibria. Levy then steps back to study the equilibrium binding agreements that characterize party *formation*. A more precise description now follows.

The set of feasible policies is some compact, convex subset Q of Euclidean space. There are n groups of voters, each group consisting of a continuum. Voters in group i share the same preferences, given by a continuous, single-peaked, strictly concave utility function $u(q, i)$ on Q.

Each group is represented by a single politician, who shares the same preferences as her group compatriots. She derives utility only from the final policy implemented, and not from the fact of winning or losing.[5]

First we describe coalitional equilibrium. Take as given a party structure, which is to be interpreted as a partition of the set of all

---

[5]The case in which winning carries an additional payoff is another possible application. The specific results will be different, but the same methodology applies.

politicians into coalitions ("parties"). The parties contest a general election. Each party can announce a platform $q \in Q$, or some "null platform" $\emptyset$ which may be interpreted as not running. With some abuse of notation denote by $\mathbf{q}$ the full vector of platforms (including the nulls if any). It has dimensionality equal to the number of parties.

The main restriction on announced party platforms is that they must have ex post credibility. To describe this, define the *Pareto set* (of policies) for coalition S by

$$Q(S) = \{q \in Q | \text{ For no } q' \in Q \text{ is } \mathbf{u}(q', S) \gneq \mathbf{u}(q, S)\}.$$

Party S must announce a policy from $Q(S)$. No other announcement is credible, as it will be presumed S will immediately renegotiate its platform to some location in $Q(S)$, if it wins. In short, the action set for S is $Q(S) \cup \emptyset$.[6]

Now define payoffs. Assume that if all platforms are null the payoff to each politician is lower than the minimum value of all u's over all policies. Otherwise, each voter is assumed to vote sincerely for her favorite policy from $\mathbf{q}$, and voters who are indifferent over one or more policies split their votes uniformly. Write $W(\mathbf{q})$ for the subset of platforms with the highest vote shares, and let $w(\mathbf{q})$ stand for the cardinality of $W(\mathbf{q})$.

The winning policy is determined by plurality rule: a policy is chosen with equal probability from $W(\mathbf{q})$. So the expected utility to politician i from the vector of platforms $\mathbf{q}$ (not all null) is just

$$u_i(\mathbf{q}) \equiv \frac{1}{w(\mathbf{q})} \sum_{q \in W(\mathbf{q})} u(q, i).$$

We complete the description of coalitional equilibrium. Subject to the credibility constraint imposed by $Q(S)$, each party S best-responds to the actions of others by choosing a Pareto-optimal

---

[6]At first glance, then, this formalism does not fit precisely with our description of a coalitional game, in which an action vector was chosen from the product of all member action sets. But this is easily dealt with. For instance, define a "renegotiation mapping" which takes all policy profiles in the member product set to a single element of $S(S)$. When the former profile is announced, voters know that the final outcome will be renegotiated using the renegotiation mapping. This yields exactly the same game (but with a lot of extra unhealthy notation), and the fit with our general framework is perfect.

action, as measured by the utilities $\mathbf{u}_S(\mathbf{q})$. A coalitional equilibrium is the joint outcome of such best responses.

Because it is better to have any policy than no policy at all, so a coalitional equilibrium will always be associated with a non-null policy vector $\mathbf{q}$. This describes a mapping from party structures π to the set of all coalitional equilibria $\beta(\pi)$.

We can now study the equilibrium party structures and policies generated by EBAs, starting from the grand coalition of all parties. In the interest of space, I avoid a general analysis, but borrow two examples from Levy (2004).

EXAMPLE 12.1. *There are three groups (and politicians), so $N = \{123\}$. No group has a majority. The policy space is $Q = [1,3]$. Group i's preferences are given by $u(q,i) = -(i - q)^2$, for $i \in N$.*

In this example, group 2 contains the median voter. Therefore, under the party structure of singletons, the unique coalitional equilibrium outcome involves the implementation of $q = 2$.

When the party structure is given by $\{\{12\}, \{3\}\}$, there is a continuum of coalitional equilibria, in which any policy $q \in (1,2]$ can be implemented. Similarly, under $\{\{1\}, \{23\}\}$, any policy in $[2,3)$ is a coalitional equilibrium outcome. But it is obvious that only one equilibrium outcome in each set can be an EBA: that involving the median policy $q = 2$. Politician 2 can block any other outcome by perpetrating the singletons.

What about the partition $\{\{13\}, \{2\}\}$? In this case, there is a unique coalitional equilibrium outcome with policy $q = 2$. One implementation of this outcome is that party $\{13\}$ stays out while 2 chooses $q = 2$. There is actually another implementation in which 2 stays out while $\{13\}$ jointly choose $q = 2$! In any case, only the median voter's policy can be implemented so that these coalitional equilibria are also EBAs.

These observations prove that the only EBA for the grand coalition of all parties involves the median voter's policy as well. Politician 2 can perpetrate the structure $\{\{13\}, \{2\}\}$ and block any other outcome for that coalition.

In this example, the ability to form parties and write equilibrium binding agreements gets us nowhere. The equilibrium *outcome* is unchanged at the median voter's policy.

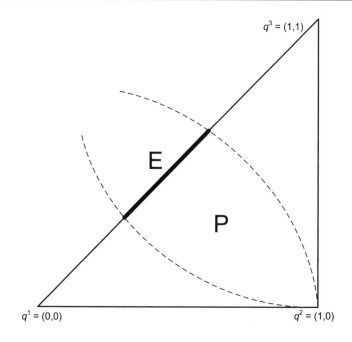

FIGURE 12.1. POLICY SPACE FOR EXAMPLE 12.2

Matters are different, however, when the policy space is multidimensional.

EXAMPLE 12.2. *There are three groups (and politicians), so N = {123}. No group has a majority. The policy space Q has two dimensions, x and y. It is the triangle formed by the three ideal points $q^1 = (q^1_x, q^1_y) = (0,0)$, $q^2 = (q^2_x, q^2_y) = (1,0)$ and $q^3 = (q^3_x, q^3_y) = (1,1)$. Group i's preferences are given by $u(q,i) = -\alpha(q^i_x - q)^2 - (1 - \alpha)(q^i_y - q)^2$, for $i \in N$, where $0 < \alpha < 1$.*

Figure 12.1 illustrates the policy space and ideal points. It is obvious (given $\alpha \in (0,1)$) that voters in groups 1 and 3 feel closer to the ideal point of group 2 than they do to those of each other. Everything else is symmetric, so without loss of generality we additionally restrict α to be no greater than $1/2$. This implies that group 2 (weakly) prefers the ideal point of group 1 to that of group 3.

Levy interprets this multidimensional environment as follows: Think of groups 2 and 3 as consisting of low-income voters; while group 1 contains high-income voters. The x-axis may then be viewed as a tax rate, so that the high-income voters ideally have a zero tax ($q_x^1 = 0$), while the others prefer full expropriation ($q_x^j = 1$ for $j = 2, 3$).

Now think of expenditures on the y-axis as pure waste as far as groups 1 and 2 are concerned, but catering to the needs of the "special interest" group 3. To be concrete, suppose that when $y = 0$, all tax revenues are spent on general public goods, and when $y = 1$, they are all spent on some local public good that benefits group 3 alone.

To study EBAs, begin with coalitional equilibria for the singletons. Despite the multidimensional nature of the situation, the policy $q^2 = (1, 0)$ is the unique coalitional equilibrium outcome, and it is an EBA. In our interpretation: when all politicians run separately, there are high taxes in equilibrium and all the revenues are spent on public goods.

Observe that there are policies that groups 1 and 3 jointly prefer to q^2: Figure 12.1 shows that these lie within the area marked P, between the two indifference curves of groups 1 and 3 that pass through q^2. However, when politicians from groups 1 and 3 run on their own, they cannot commit to offer such platforms.

When the party structure is given by $\{\{12\}, 3\}$ (or by $\{1, \{23\}\}$), there is, as in Example 12.1, a continuum of coalitional equilibria, but only one of these is an EBA: the one that implements q^2. I leave the simple details of the analysis to the reader.

The partition $\{\{13\}, 2\}$ is, however, of particular importance. Observe that the Pareto set for the coalition $\{13\}$ is

$$Q(\{13\}) = \{(z, z) | z \in [0, 1]\},$$

the line segment joining q^1 and q^3. As shown in Figure 12.1, the darkened portion of this line segment that intersects with P contains policies that would win against q^2 in a binary contest; all other policies will result in 2 running, winning, and implementing q^2. Therefore, using the specific forms of the utility function that we've chosen for the example, an equilibrium for this party structure must involve the implementation of some policy of the form (z, z), where $z \in (1 - \sqrt{1 - \alpha}, \sqrt{\alpha})$.

Indeed, each of these outcomes is also an EBA. Call this set of outcomes E.

We may now describe EBAs starting from the grand coalition. Consider any outcome in the Pareto set of {123}. Because politicians 1 and 3 can individually depart and assure themselves the outcome q^2, an EBA for the grand coalition must lie in the set P. At the same time, {13} can block an outcome using an EBA from the set E. Therefore an EBA for the grand coalition *must* come from E as well. Finally, note that politician 2 can also precipitate the coalition structure {{13}, 2}. The notion of optimism inherent in our definition therefore pins down a unique EBA for the game as a whole: the outcome that maximizes group 2's payoff among the set E.[7]

In short, when there are multidimensional platforms, two groups may get together and form a coalition that outwits a third group. Moreover, under the interpretation that we have adopted, it is clear that *any* two groups will not do. In our example, it is the high-income group that joins hands with the special interest group. The former group accepts the inevitability of some taxation, but manages to avoid outright expropriation, which would be the outcome in a world without coalitions. The reason they manage to avoid expropriation is that they agree, moreover, to have the tax revenues used (in part) for special-interest activities. Together, they are in a position to out-maneuver the "median voter", who would otherwise have achieved far-reaching redistribution with no concessions to special-interest groups.

Levy (2004) shows how these examples can be translated into general results. For example, her Proposition 3 (p. 265) establishes sufficient conditions for EBAs to generate outcomes that differ from the "median-voter" outcomes of singleton coalition structures. As in the examples, the multidimensionality of the underlying policy space plays a crucial role.

We see, then, that equilibrium binding agreements may be an effective tool for the study of political coalition formation. Ray and Vohra (1997) and Yi (1996) contain other applications.

---

[7]Without the assumption of optimism, one could obtain the entire set E as the collection of EBAs for the grand coalition. This changes none of the interpretation or discussion, however.

12.5 Beyond Nested Deviations

A central assumption underlies the concept of equilibrium binding agreements, which is that all threats to an existing coalition structure come "from within". That is, only subsets of existing coalitions can block a proposed agreement. This is a perfectly reasonable assumption in many situations. It may well be that members of a coalition, once formed, do not (or can not) communicate with individuals in some other coalition. Each coalition writes its own binding agreements, and if there are objections raised to certain potential agreements, it is natural to suppose that these come entirely from within the coalition.

Reasonableness apart, the assumption of nested coalitional threats possesses another advantage. Equilibrium coalition structures may then be recursively defined. We did just this, starting from the singletons. (A similar restriction may be found in coalitional refinements of noncooperative equilibrium, such as the coalition-proof Nash equilibrium of Bernheim, Peleg and Whinston (1987).)

That said, there are obviously many other situations in which fresh coalitions can form more freely, and we would like to extend the theory to encompass such situations. There are two steps here. First, one might allow arbitrary coalitions to come together, starting from any coalition structure, insisting nonetheless that all agreements are irreversible. This step would bring the blocking theory roughly into line with the irreversible commitments model based on bargaining (and covered in Chapters 4–7). Second, one might allow for ongoing negotiations, as coalitions form and re-form, in the spirit of Chapters 8–10. In this section, I briefly discuss how one might take the first step, leaving the second to the next chapter.

I base the discussion that follows on Diamantoudi and Xue (2007).

The main problem with constructing a definition of equilibrium binding agreements for the case of arbitrary coalitional deviations is one of "circularity". We would like a possible agreement to be blocked, if at all, only by an equilibrium binding agreement. That, in turn, relies on foreknowledge of which agreements are binding "to begin with". When coalitional deviations are nested, this problem does not arise as we can proceed recursively.

One way to deal with the general issue is to think of equilibrium binding agreements as a fixed point of an appropriately defined blocking relation. In words, that fixed point would (loosely) declare: "The set of equilibrium binding agreements is the collection of all agreements that are 'unblocked' by the set of equilibrium binding agreements." As Greenberg (1990) and Chwe (1994) and several others have noted, that fixed-point role is often admirably played by the *stable set* of von-Neumann and Morgenstern. This is how Diamantoudi and Xue (2007) proceed to extend the concept of an EBA.

The first line of business is to settle upon a suitable blocking relationship. To do so, we return to the notion of sequential blocking in Section 12.2.1. The advantage of this notion is that it specifies a single, farsighted "chain" along which perpetrators move in sequence.[8] We will need to widen the notion to allow for arbitrary coalitions to form. This is very easy to do.

Begin with $(\pi, \mathbf{a})$. Let S be any coalition, to be regarded in the sequel as a perpetrator (S is not necessarily a subset of some member of π). Say that coalition S *induces* $(\pi', \mathbf{a}')$ from $(\pi, \mathbf{a})$ if

1. Some partition of S is included in the new structure π'.

2. The residual (after S breaks off) of any coalition in π belongs in π'.

3. All unaffected coalitions in π continue to belong in π'.

4. The profile $\mathbf{a}'$ is a coalitional equilibrium under π'.

Condition 1 allows for the perpetrator S to organize itself, if it so pleases, into subgroups.[9] Under a different interpretation, this allows for two or more perpetrating coalitions to simultaneously block an allocation (view S as the union of these perpetrators, and allow each of them to separately belong to π'). Conditions 2 and 3 are perfectly self-explanatory; I add here only a formal description that incorporates both: if coalition $T \in \pi$, then $T - S \in \pi'$.

---

[8]One might also adapt — at the cost of some additional complexity — the blocking notion used here.

[9]This makes no difference to the connection between my original notion of blocking and sequential blocking, explored in Section 12.2.1.

Finally, condition 4 requires the new payoff vector to be a coalitional equilibrium payoff under the new structure π'.[10]

We may now define sequential blocking in a manner analogous to Section 12.2.1. Say that $(\pi', \mathbf{a}')$ *sequentially blocks* $(\pi, \mathbf{a})$ if there exists a sequence $\{(\pi^0, \mathbf{a}^0), (\pi^1, \mathbf{a}^1), \ldots, (\pi^m, \mathbf{a}^m)\}$ such that:

[S.1'] $(\pi^0, \mathbf{a}^0) = (\pi, \mathbf{a})$, $(\pi^m, \mathbf{a}^m) = (\pi', \mathbf{a}')$ and for every $i = 1, \ldots, m$, there is a coalition S^i that induces $(\pi^i, \mathbf{a}^i)$ from $(\pi^{i-1}, \mathbf{a}^{i-1})$.

[S.2'] $\mathbf{u}_{S^i}(\mathbf{a}') \gg \mathbf{u}_{S^i}(\mathbf{a}^{i-1})$ for all $i = 1, \ldots, m$.

We may now complete the definition. To each coalition structure π assign a set $\mathcal{B}^*(\pi)$ of allocations; call them *extended equilibrium binding agreements* (EEBA) following Diamantoudi and Xue (2007). To be sure, for some coalition structures such a set may be empty. The defining property of this assignment is the provision that for every π, $\mathcal{B}^*(\pi)$ is precisely the set

$$\{\mathbf{a} \in \beta(\pi)| \text{ no } (\pi', \mathbf{a}'), \text{ with } \mathbf{a}' \in \mathcal{B}^*(\pi'), \text{ sequentially blocks } (\pi, \mathbf{a})\}.$$

Notice the circularity of this description: $\mathcal{B}^*$ appears within its own "definition". What this really is, then, is a consistency condition which any assignment of EEBAs to coalition structures must satisfy. Indeed, there may be several such consistent assignments, and there may well be none.

The reader familiar with cooperative game theory will see in this condition a restatement of the von-Neumann–Morgenstern stable set. That set may be defined for any abstract binary relation $>$ on any abstract set X. A subset Y of X is *internally stable* if for no two elements z and w of Y is it the case that $z > w$. And Y is *externally stable* if for every z not in Y, there is w in Y such that $w > z$. A set is *stable* if it is both internally and externally stable.

Observe that internal stability can be dealt with by taking Y to be any singleton set, while external stability can be assured by taking $Y = X$. The tension behind the concept lies in assuring both internal and external stability within one set.

Internal and external stability may be combined as follows: say that Y is *stable* if Y is the collection of *all* elements of X that are unblocked

---

[10]Notice how this implicitly presumes that coalition S can induce *any* such payoff. Those uncomfortable with this may simply wish to assume that each coalition structure is associated with a unique coalitional equilibrium payoff vector.

using elements of Y. This statement is equivalent to our definition of EEBA, which is why EEBA corresponds to a stable set.[11]

The relationship between EEBAs and EBAs is subtler than appears at first sight. Allowing for all kinds of coalitions — not just internal threats — means that *more* blocks are potentially available. This suggests that *less* allocations can stand the blocking test, so the set of equilibrium binding agreements must shrink. This suggests that EEBAs should be a subset of the set of EBAs. The problem with this kind of reasoning is that it neglects the fact that blocking allocations must pass the credibility test. Therefore the very forces that make blocking easier also serve to make *credible* blocking harder. Therefore, no simple inclusion relationship exists between equilibrium binding agreements and their extended counterparts.

In particular, there are situations in which some EEBA assignment may be efficient, while other EEBA assignments are more closely related to EBAs. As an example, recall the game used to prove the inefficiency result for EBAs in Proposition 12.1. I reproduce the matrices here for easy reference.

		x_{2a}	x_{2b}	x_{2c}
x_{3a}	x_{1a}	2.6, 2.6, 2.6	3.2, 2.2, 3.2	3.7, 1.7, 3.7
	x_{1b}	2.2, 3.2, 3.2	2.7, 2.7, 3.7	3.1, 2.1, 4.1
	x_{1c}	1.7, 3.7, 3.7	2.1, 3.1, 4.1	2.6, 2.6, 4.6

		x_{2a}	x_{2b}	x_{2c}
x_{3b}	x_{1a}	3.2, 3.2, 2.2	3.7, 2.7, 2.7	4.1, 2.1, 3.1
	x_{1b}	2.7, 3.7, 2.7	3.1, 3.1, 3.1	3.6, 2.6, 3.6
	x_{1c}	2.1, 4.1, 3.1	2.6, 3.6, 3.6	2.9, 2.9, 3.9

		x_{2a}	x_{2b}	x_{2c}
x_{3c}	x_{1a}	3.7, 3.7, 1.7	4.1, 3.1, 2.1	4.6, 2.6, 2.6
	x_{1b}	3.1, 4.1, 2.1	3.6, 3.6, 2.6	3.9, 2.9, 2.9
	x_{1c}	2.6, 4.6, 2.6	2.9, 3.9, 2.9	3.3, 3.3, 3.3

OBSERVATION 12.2. *There is an EEBA assignment for this game, in which $\mathcal{B}^*(\pi)$ is empty for every coalition structure except for the structure $\pi^G = \{N\}$, and $\mathcal{B}^*(\pi^G) = \{(3.3, 3.3, 3.3)\}$. The unique binding agreement under this assignment is efficient.*

---

[11]As Diamantoudi and Xue (2007) show, the EBA notion is connected to another stable set, one which is based entirely on the internal blocking relation.

There is another EEBA assignment consisting of the three asymmetric coalition structures and their associated coalitional equilibrium payoffs. These binding agreements are inefficient.

There is no other EEBA assignment.

Proof. To establish this claim, observe that the first specified assignment is indeed an EEBA. It suffices to show that every $(\pi, \mathbf{a})$ with $\mathbf{a} \in \beta(\pi)$ is sequentially blocked by $(\pi^G, (3.3, 3.3, 3.3))$. The singleton structure (with associated payoff $(2.6, 2.6, 2.6)$) is sequentially blocked in a single step: use N as the blocking coalition leading to π^G. The asymmetric structure $\{\{1\}, \{23\}\}$ with unique associated payoff vector $(3.7, 2.7, 2.7)$ is sequentially blocked in two steps: first player 2 precipitates the singletons, and then N moves. By symmetry, we are done.

As for the second assignment, observe that no asymmetric structure can sequentially block another asymmetric structure, so internal stability is guaranteed. Both the outcome $(\pi^G, (3.3, 3.3, 3.3))$ as well as $(\pi^*, (2.6, 2.6, 2.6))$ are sequentially blocked (by each of the asymmetric structures), which verifies external stability.

Now, no EEBA assignment can contain the singleton structure π^*. So any other assignment must have either the grand coalition structure or one (or more) of the asymmetric structures. If the former, then no asymmetric structure can be included, for internal stability would be violated otherwise. This describes our first assignment. If the latter, then *all three* of the asymmetric structures must be included, otherwise external stability would be violated. This is our second assignment, and so the observation is verified.   □

Observation 12.2 illustrates a number of features. To begin with, the necessary circularity of the EEBA construction often results in the existence of several stable assignments. The observation describes two assignments. Of course, such multiplicity is not peculiar to EEBAs. It is well-known that the very same underlying situation may admit several von-Neumann–Morgenstern stable sets.

That doesn't mean, of course, that "anything can happen". For instance, in the example above, the singleton structure can never be part of an EEBA assignment.

The same bootstrapping that generates multiplicity might also jeopardize existence. Starting with the classic counterexample of

Lucas (1968), it is known that stable sets do not, *in general*, exist. Whether such examples of nonexistence can be found for EEBAs is currently an open question.

Finally, the observation reveals the existence of an efficient EEBA in the example. One might wonder if this is a general finding. Diamantoudi and Xue (2007) discuss this issue in some detail. Their Proposition 2 may be viewed as a generalization of the findings in Observation 12.2, laying down sufficient conditions for the existence of an efficient EEBA assignment. At the same time, they provide an example in which there is *no* efficient EEBA assignment. That inefficiency may persist even when all agreements are binding is a central theme of this monograph. As Diamantoudi and Xue (2007) observe, their example "reinforces the inefficiency puzzle".

12.6 Summary

This chapter studies irreversibly binding agreements using a traditional methodology based on blocking, so familiar from cooperative game theory. As described in Chapter 11, the blocking notion needs to be amended substantially to allow for both farsighted behavior as well as externalities. We begin by introducing the concept of *equilibrium binding agreements*. These are agreements that cannot be blocked — with the farsighted implications factored in — by *other* equilibrium binding agreements. The obvious circularity of the definition is cut through by presuming that only *subsets* of existing coalitions can deviate, thus permitting a recursive construction.

Using the concept of an equilibrium binding agreement, we return to the ever-present question of efficiency. We show that there is a large class of games for which equilibrium outcomes indeed fail to be efficient, despite the ability to write binding agreements. This observation reinforces similar findings for the coalitional bargaining model studied in Part 2.

Next, we study an application of the solution concept to political party formation. We argue that the concept is useful in predicting certain types of political coalitions.

The assumption that only subcoalitions of existing coalitions can block existing allocations may be useful in many situations, but it is restrictive. In the final part of this chapter we attempt to relax this assumption, and in so doing connect our extended concept to other classical and recent notions in cooperative game theory.

An exploration of continued negotiations, in which all commitments are reversible, is the subject of the next chapter.

Appendix

Proof of Observation 12.1. Suppose $(\pi', \mathbf{a}')$ blocks $(\pi, \mathbf{a})$. Let S^1 be the leading perpetrator in this move. If there are no other perpetrators, then it is clear that $(\pi', \mathbf{a}')$ sequentially blocks $(\pi, \mathbf{a})$ with $m = 1$. Suppose, therefore, that the set of other perpetrators is $\mathcal{T} = \{S^2, \ldots, S^m\}$. Define π^1 to be the coalition structure obtained by re-merging all other perpetrators in $\mathcal{T}$. Since $(\pi', \mathbf{a}')$ blocks $(\pi, \mathbf{a})$, by condition [E.3], $\mathcal{B}(\pi^1) = \emptyset$, and there exists $\mathbf{a}^1 \in \beta(\pi^1)$ such that $(\pi', \mathbf{a}')$ blocks $(\pi^1, \mathbf{a}^1)$, with a leading perpetrator from $\mathcal{T}$. Without loss of generality let this leading perpetrator be S^2. Define π^2 to be the coalition structure obtained by re-merging all perpetrators $S^3, \ldots, S^m$ with their respective residuals. By appealing to Condition E.3 yet again, we can assert that there exists $\mathbf{a}^2 \in \beta(\pi^2)$ such that $(\pi^2, \mathbf{a}^2)$ is blocked by $(\pi', \mathbf{a}')$ with S^3 (say) as a leading perpetrator. In this way we obtain a sequence $(\pi, \mathbf{a}), (\pi^1, \mathbf{a}^1), \ldots, (\pi^m, \mathbf{a}^m)$ such that (a) $(\pi^m, \mathbf{a}^m) = (\pi', \mathbf{a}')$, (b) for every $i = 1, \ldots, m$, $\mathbf{a}^i \in \beta(\pi^i)$, and (c) for every $i = 1, \ldots, m$, $(\pi^{i-1}, \mathbf{a}^{i-1})$ is blocked by $(\pi', \mathbf{a}')$ with S^i as the leading perpetrator. Clearly, $\{(\pi, \mathbf{a}), (\pi^1, \mathbf{a}^1), \ldots, (\pi^m, \mathbf{a}^m)\}$ satisfies conditions [S.1] and [S.2]. □

CHAPTER 13

The Blocking Approach in Real Time

13.1 Introduction

The motivation for this chapter is exactly the same as that for Chapters 8–10. There are numerous situations in which the commitment to form a coalition is irreversibly binding. But there are many others in which negotiations can, in principle, open and reopen over an indefinite period of time (see the introduction to Chapter 8 for a detailed discussion of such issues). To study this phenomenon requires an explicitly dynamic model of coalition formation.

What I'd like to do in this chapter, which is based on Konishi and Ray (2003), is lay down a model of coalition formation in real time that runs parallel to the setup in Chapter 8. The main difference between the two approaches is that while the earlier model takes an explicit bargaining-theoretic viewpoint, the current model is faithful to the "blocking" methodology of traditional cooperative game theory. As we shall see, there is far less of an emphasis here on precise protocols and individual strategies, and more a transplantation of coalitional blocking notions to a dynamic environment.[1]

There is a second difference between the analysis of these earlier chapters and the one to be conducted here. In Chapter 8, I was careful to restrict the analysis to binding agreements, in which additional changes needed to be approved by those individuals who were "directly affected" by those changes. The notion of what

---

[1]I use the terms "earlier" and "current" models to refer to their order of appearance in this monograph. The correct chronology is that the Konishi–Ray framework in this chapter was developed several years before the bargaining variant studied in Chapters 8–10.

it meant to be "directly affected" was not straightforward (recall the conditions B.1 and B.2 from Section 8.3) but it certainly included alterations in coalitional membership.

I could do the same in this chapter. Indeed, there are probably pedagogical gains to retracing old results with a different methodology. But I felt it more useful (especially in a short monograph) to put that new methodology to work on a different —though complementary — set of issues. Accordingly, I am changing not only the methodology, but will often emphasize *temporary agreements* for a change. These are agreements which are also binding, but only for a single period at a time. I will attempt to capture the temporary nature of such agreements by suitably defining approval committees for various changes in state; see below for details.

13.2 An Informal Description

Let N be a set of players and X a set of states. Suppose that for each state in X and each coalition S, a possible set of "coalitional moves" (by S) to some subset of states is given. A mapping from the current state to a probability distribution over the set of all coalitional moves feasible at that state induces a dynamic process on X. Noting that moves are associated with actions taken by coalitions, we call this a *process of coalition formation*.

Under such a process players receive (additive discounted) utility from the entire path of states. This induces a value for each player in the standard way, as a function of the going state.

This framework is very similar to the one used in Chapters 8–10, except that there is no protocol, no proposals or counterproposals, and no accept-reject decisions. Instead, as we now proceed to describe, we use *coalitions* as our fundamental behavioral units.

A process of coalition formation is an equilibrium if at any date and at any going state, a coalitional move to some other state can be "justified" by the very same scheme applied in future: the coalition that moves must have higher present value (starting from the state it moves *to*) for each of its members, compared to (one-period) inaction under the going state.

In the most general form that we study it, a process of coalition formation precipitates a Markov process on X, the uncertainty reflecting both the choice of the deviating coalition at some state (there may be several potential deviants) and the choice of state that the coalition deviates to (there may be several potential moves). The use of value functions induced by the scheme itself implies perfect foresight on the part of all coalitions: players expect and understand that coalitions may move in the future, and form (common) beliefs about the likelihood of such events.

The explicitly dynamic nature of our definition possesses at least three advantages relative to existing formulations.

First, by allowing all moves to take place in real time, as it were, the definition allows us to bridge the gap between myopic notions of stability (such as those implicit in the core or the bargaining set) and the more recent definitions based on farsightedness (such as those discussed in the previous chapter). Extreme myopia would correspond to a discount factor of zero, while extreme farsightedness would be approximated as the discount factor converges to unity.

The point is that the static concepts based on farsightedness are really attempting to capture a fundamentally dynamic process, in which an action may generate an entire chain of reactions. My formulation takes this dynamic story seriously instead of writing down a shorthand for it.

Second, the theory of coalitional deviations based on blocking is often made complicated by the issue of multiple continuations following a single deviation. In the previous chapter we discussed this point in the context of EBAS (but see also Greenberg(1990), Chwe (1994), Xue (1998) and many others). Greenberg distinguishes between optimistic and conservative "standards of behavior", in which currently deviating coalitions evaluate the future multiplicity of other deviations in hopeful or pessimistic ways. This approach to the treatment of multiplicity can be avoided by borrowing more freely from the language of repeated or dynamic games, which we do. Future paths (perhaps probabilistic in nature) are evaluated using common beliefs (as embodied in the transition probability) and expected payoffs are calculated using these beliefs. In particular, the problem of predicting how some complementary coalition might respond to a deviation (see Section 12.2.2) is automatically taken care of, albeit through the use of a specific solution concept.

Third, several solution concepts, especially those that concern themselves with farsighted agents, inevitably run into the problem of cycles (for an early discussion of this, see Shenoy (1979)). Chains of coalitions may appear and reappear in the blocking process.[2] In the approach adopted in this chapter, recurrent cycles of moves pose no problem at all. Payoffs from such cycles are simply to be evaluated as any sequence of payoffs is evaluated: by adding up discounted one-period returns over time.

A particularly relevant interpretation of cyclical outcomes — one which forms the principal motivation for this chapter — arises from the possibility of constant renegotiation. Agreements may be torn up and rewritten, especially if the environment external to a particular coalition is altered by the formation of other coalitions (note that this would be irrelevant for characteristic functions, but especially important when there are widespread externalities).

13.3 A Process of Coalition Formation

13.3.1 Basic Ingredients. Let N be a finite set of *players* and X a finite set of *states*.[3] A *coalition* is a nonempty subset S of N. For each state x in X and each coalition S, define $F_S(x)$ to be the set of states achievable by a one-step coalitional move (by S) from x. That is, for every state $y \in F_S(x)$, S is a valid approval committee for the move from x to y, just as in Chapter 8. A coalition S always has the option to do nothing, so x is always a member of $F_S(x)$.

If player i is equipped with a one-period von Neumann-Morgenstern payoff function u_i on X and a discount factor $\delta_i \in (0,1)$, her payoff from a sequence of probabilities $\{\lambda_t\}$ on X may be written as

$$\sum_{t=0}^{\infty} \delta_i^t \left(\sum_{x \in X} \lambda_t(x) u_i(x) \right).$$

---

[2] One approach is to exclude cycles explicitly by assuming the nestedness of coalitional moves, as in the previous chapter (see also Bernheim, Peleg and Whinston (1987) and Ray (1989)). Alternatively, one might exclude cycles by implicitly assuming that such cycles gives the worst payoffs (see, e.g., Mariotti (1997) or Xue (1998)). Finally, one might study coalition formation — as we've done earlier in this book — within a bargaining context, in which infinite bargaining delays result in zero payoff.

[3] The restriction to a finite set of states is for technical convenience. For instance, I do not know whether existence results such as Proposition 13.1 below extend to the general infinite case, though an extension to a countable infinity of states is fairly standard.

13.3.2 Interpreting States. Several interpretations of a "state" are possible. In Chapter 8, we viewed a state as a pair: a coalition structure, coupled with a vector of payoffs under a coalitional equilibrium for that structure. While this continues to be a leading interpretation in this chapter, there are others. For instance, a state could represent a profile of actions taken in some normal-form game. As discussed in the Introduction to this chapter, our main innovation will be to study temporary agreements, and so the specification of F_S is rather more important. Here are some examples.

1. *A Characteristic Function.* Consider the simplest two-person NTU characteristic function, in which there are two coalition structures with a single payoff vector in each. Let x_1 be the state with structure {12}, and x_2 the state with singleton structure. Then for $x \in \{x_1, x_2\}$, u_i is just the payoff to player i under the corresponding structure.

If all agreements are permanently binding (in the sense of Chapter 8), then all that a single individual can approve is the going state: $F_{\{i\}}(x) = \{x\}$, while the grand coalition can additionally precipitate x_1 from x_2 (and vice versa). If agreements are temporary then each singleton coalition can additionally approve x_2 from x_1 (but not x_1 from x_2 unilaterally). In this chapter we will be particularly interested in the latter formulation.

2. *A Partition Function.* Suppose that there are three players. If all three stand together, the payoff vector is (a, a, a). If all stand alone, the payoff vector is $(0, 0, 0)$. If i is alone and j and k are together, the payoff is b to i and c each to the other two. These determine the functions $u_i(x)$ for each player i and and each state x.

Once again, much of the description of F_S is obvious. But partition functions pose new issues. Suppose that agreements are temporary, and i moves away from the grand coalition. Is the resulting structure $\{i\}, \{jk\}$ or the singleton structure $\{\{i\}, \{j\}, \{k\}\}$? This time it is more than a mere question of interpretation, and the dynamic model of coalition formation just described forces us to take a stand on the matter. So far as the formal theory is concerned, this is not an issue as long as $F_S(x)$ is fully specified for all coalitions S and states x.

3. *A Game in Strategic Form.* A situation in which a normal form game is played at every date is particularly easy to embed. Let N be a set of players, and let A_i be the (finite) action set of player i. A state is simply an action profile $\mathbf{a} = (a_i)_{i \in N}$. Starting from some

action profile **a** a coalition S can induce any action profile of the form $(\mathbf{a}'_S, \mathbf{a}_{-S})$, where $\mathbf{a}_{-S}$ is that part of the profile not chosen by members of S, and $\mathbf{a}'_S$ is any vector of actions on the part of S.

Such an interpretation, while simple, is not devoid of conceptual issues. Why is $\mathbf{a}_{-S}$ fixed when S moves? One interpretation is that an action profile constitutes a temporarily binding agreement, and at every date some coalition receives the opportunity to costlessly renege on such an agreement.

13.3.3 Equilibrium. A *process of coalition formation* (PCF) is a transition probability $p : X \times X \rightarrow [0, 1]$ (so that $\sum_{y \in X} p(x, y) = 1$ for each $x \in X$).

We interpret p as capturing the (possibly stochastic) transitions from one state to another. These transitions will be induced by coalitions who stand to benefit from them (see below). The restriction that the PCF be Markov means that the corresponding "coalitional strategies" are Markov too.[4]

A PCF p induces a *value function* V_i for each player i. This value function captures the infinite horizon payoff to a player starting from any state x, under the Markov process p. Standard observations tell us that the value function for i must be the unique solution to the functional equation

(13.1) $$V_i(x, p) = (1 - \delta_i)u_i(x) + \delta_i \sum_{y \in X} p(x, y)V_i(y, p).$$

We are now in a position to define a central concept in this chapter, the notion of "profitable moves". These will be used to impose restrictions on an equilibrium process of coalition formation. Fix a PCF p, a state x, and a coalition S. Say that S has a *(weakly) profitable move* from x (under p) if there is $y \in F_S(x)$ (with $y \neq x$) such that $V_i(y, p) \geq V_i(x, p)$ for all $i \in S$. S has a *strictly profitable move* from x if there is $y \in F_S(x)$ such that $V_i(y, p) > V_i(x, p)$ for all $i \in S$. Finally, say that a move y is *efficient* for S if there is no other move for S, say z, such that $V_i(z, p) > V_i(y, p)$ for all $i \in S$.

A PCF is an *equilibrium process of coalition formation* (EPCF) if (i) whenever $p(x, y) > 0$ for some $y \neq x$, then there is S such that y is a

---

[4]I would like to focus on some different issues in this chapter, and so I leave a more general model with history-dependent strategies unexplored here.

(weakly) profitable and efficient move for S from x, and (ii) if there is a strictly profitable move from x, then $p(x, x) = 0$ and there is a strictly profitable and efficient move y with $p(x, y) > 0$.

Thus a going state is allowed to move to another state *only if* there is a coalition whose members all agree to move to the new state and cannot find any strictly better alternative state (under the going value functions). Moreover, if there is a strictly profitable move, then the state *must* change, and there must be at least one move to a state which is interpretable as a strictly profitable and efficient move for some coalition.

Notice that this definition allows for (but does not insist on) possible changes in state in which the initiating coalition is indifferent to the change. At the same time, the definition does not insist that *every* strictly profitable move (under the equilibrium PCF) be given positive probability. This is true in a particularly stark way of "deterministic" PCFs — to be introduced in Section 13.4 — in which only one coalition is selected to act at each state containing some profitable move (even though, in principle, there may be several such moves).

Some more observations on "efficient moves" are to be found in Appendix A.

Two more remarks assist in understanding our equilibrium concept. The first is semantic: I use the term "moves" for what would ordinarily pass as coalitional *deviations* or acts of blocking. In the traditional theory, these are off-equilibrium constructs against with the stability of a solution concept is tested. Not so here: the solution allows for coalitions to "deviate" — and keep "deviating" — in a real time definition of equilibrium. Second, the definition of equilibrium reflects the looseness of underlying protocol: even the "choice" of who moves (or deviates) at a particular state is left to the equilibrium construct, provided, of course, that such a move is justified in its profitability and efficiency.

It is this eclecticism regarding coalitional moves which permits an EPCF to always exist:

PROPOSITION 13.1. *An equilibrium process of coalition formation exists.*

While the argument relies on a familiar fixed point theorem, the construction of the fixed point map is not entirely standard. The reader interested in the details is invited to consult Appendix B.

The proposition extends to state spaces that are countable. Whether existence holds in more general cases is an open question.

13.4 Deterministic Equilibrium Processes

In this section, we narrow our definition considerably by concentrating on EPCFs with absorbing limit states and fully deterministic paths. We then relate our narrowed definition to well-known concepts in cooperative game theory. This situates the proposed solution in the context of existing literature.

A state x is *absorbing* under a PCF p if $p(x, x) = 1$. In turn, a PCF is *absorbing* if for each state y, there is some absorbing state x with $p^{(k)}(y, x) > 0$ for some $k \geq 1$.[5] A PCF has a *unique limit* if it is absorbing and possesses a single absorbing state. Finally, a PCF is *deterministic* if $p(x, y)$ is either 0 or 1 for every pair of states x and y.

Our first stop is the core.

13.4.1 Deterministic EPCFs and the Core. Recall our earlier discussion of the core. At first sight, it is a myopic notion, requiring the stability of a proposed allocation to deviations or blocks by coalitions, but not examining the stability of the deviations themselves. At the same time, Observation 11.1 tells us that the the core automatically embodies a certain degree of farsightedness, insofar as "internal" chains of deviations by *nested* coalitions are concerned. We now show that each element of the core in an arbitrary characteristic function game can be "supported" (in a sense to be made precise below) as the outcome of a deterministic EPCF with unique limit. In other words, the core passes a further consistency test in which nested deviations are dispensed with.

We also establish a converse that yields an almost-complete characterization of deterministic EPCF's with unique limit.

Fix a finite set N of players. A (finite) *characteristic function* is a map $\mathbf{U}$ that associates with each coalition S a nonempty finite set of payoff vectors in $\mathbb{R}^S$. Normalize so that all payoffs are nonnegative.

---

[5]The notation $p^{(k)}$ describes the k-step transition probability derived from p in the usual way.

A *state* of a characteristic function is a pair $x = (\pi, \mathbf{u})$, where π is some partition of the player set into coalitions, and $\mathbf{u}$ is a payoff vector such that $\mathbf{u}_S \in U(S)$ for any coalition $S \in \pi$.

A *strong core state* is a state $x = (\pi, \mathbf{u})$ such that there is no coalition S and payoff $\mathbf{u}' \in U(S)$ with $\mathbf{u}' \geq \mathbf{u}_S$ and $\mathbf{u}' \neq \mathbf{u}_S$. A *weak core state* is a state $x = (\pi, \mathbf{u})$ such that there is no coalition S and payoff $\mathbf{u}' \in U(S)$ with $\mathbf{u}' \gg \mathbf{u}_S$. Obviously, a strong core state is a weak core state.

We now embed a characteristic function into an intertemporal model of coalition formation with temporary agreements. I begin with a formal account. Let X be the collection of all states of the characteristic function. For each partition π of N and each coalition S, let W denote the set of *left-behind* players $\{i \in T - S : T \in \pi, T \cap S \neq \emptyset\}$. Let $\pi_S = \{S\} \cup \{T' \in \pi : T' \cap S = \emptyset\} \cup \pi(W)$, where $\pi(W)$ is some exogenously given (but arbitrary) partition of W. (Clearly, $\pi_S = \pi$ if (and only if) $S \in \pi$.) Now define $F_S(x)$ as any collection of states $y = (\pi', \mathbf{u}')$ such that (a) $\pi' = \pi_S$, (b) $\mathbf{u}'_S \in U(S)$, and indeed, $\cup_{(\pi', \mathbf{u}', \pi) \in F_S(x)} \mathbf{u}'_S = U(S)$, (c) $\mathbf{u}'_T = \mathbf{u}_T$ for all coalitions $T \in \pi$ such that $T \cap S = \emptyset$, and (d) if $\mathbf{u}^1, \mathbf{u}^2$ satisfy $(\pi', \mathbf{u}^i) \in F_S(x)$ for $i = 1, 2$, then $\mathbf{u}^1_{-S} = \mathbf{u}^2_{-S}$.

This unwieldy formalism is easily interpreted in words: a move is available to S if the payoff vector (restricted to S) is feasible for S, if the remaining coalition structure consists of the coalitions that S left untouched and some partition of players that S left behind,[6] if the resulting payoffs to coalitions in π which have an empty intersection with S remain the same as before,[7] and if the resulting payoff vector to all non-deviant players is independent of the particular payoff vector chosen by the deviating coalition.

Observe that our description of F_S implies that agreements are temporary. *Any* coalition S is permitted to break away at any time and "implement" a payoff vector that is feasible for it. It

---

[6]As an example, Hart and Kurz (1983) consider two formulations of coalition formation games. A Δ-game considers a situation that players who are left behind by a coalitional deviation S are dissolved and each player becomes a singleton. On the other hand, a Γ-game considers a situation that each of the complementary pieces that S left behind stays together. Since we allow any regrouping of players who are left behind by S, our setting allows for both Δ- and Γ-games.

[7]Note that, in principle, several correspondences of the form $F_S(x)$ may be written down that satisfy this "independence property": our results hold for each one of them.

is worth contrasting this description with that for (permanently) binding agreements: see conditions B.1 and B.2 in Chapter 8.

The following pair of propositions provide an almost-complete characterization of deterministic EPCFs with unique absorbing limits. Every core allocation can be described as the limit of some EPCF, and every such limit is indeed a core allocation.

PROPOSITION 13.2. *Let x^* be a strong core state of a characteristic function. Then there is $\delta^* \in (0,1)$ such that for any collection of discount factors all in $(\delta^*, 1)$ and any associated intertemporal model of coalition formation with temporary agreements, there exists a deterministic EPCF defined on that model with x^* as its unique limit.*

PROPOSITION 13.3. *Fix some characteristic function. There is $\delta^* \in (0,1)$ such that for any collection of discount factors all in $(\delta^*,1)$, and for any deterministic EPCF defined on any associated intertemporal model of coalition formation with x^* as its unique limit, x^* must be a weak core state.*

For formal proofs, see Section 13.6.

The propositions show that for characteristic functions, the concept of the core and that of a deterministic EPCF with unique limit are (essentially) equivalent,[8] whenever players are sufficiently far-sighted. The limit outcomes of deterministic equilibrium processes are not only efficient, they are efficient in the strong sense that no coalition can unilaterally improve on those outcomes. And the converse is also true: for every core allocation, a suitable EPCF can be constructed with absorbing limit precisely equal to that core allocation. These results greatly extend the credibility properties of the core.

Notice how the assumption of temporarily binding agreements plays a role here. If agreements were permanently binding, as in Chapter 8, one would conjecture once again that the limit outcome must be efficient — a result analogous to Proposition 9.2. But core membership would not be obtained, simply because the existing coalition structure might preclude the formation of certain coalitions

---

[8]That is, they are equivalent up to the minor gap between strong and weak core states. To the purist interested in closing this gap: it cannot be done free of charge. Konishi and Ray (2003) provide two examples, one showing that Proposition 13.2 does not hold for all weak core states, while the conclusion of Proposition 13.3 cannot be restricted to strong core states.

(approval would be needed from others to break away). In contrast, when agreements are temporary, *any* coalition can always form, which suggests that an absorbing outcome must be impervious to all possible threats. To be sure, those threats may in turn be blocked by other counterthreats, and those counterthreats by further threats, and this is precisely what makes the proposition nontrivial. But the argument does suggest, albeit in an extremely rough way, that a limit outcome must be immune to all possible coalitional blocks.

This rough intuition actually works perfectly when individuals are extremely myopic: that is, when δ is very close to *zero* rather than unity. Then only the payoff consequences of the immediate block matter to a coalition, and it is obvious that a unique limit outcome must be a core allocation. So the propositions are actually true both when players are very myopic and when they are very farsighted. The former case represents just the traditional notion of blocking, but the latter is completely new. Indeed, in this case, a deterministic EPCF may rule out non-core allocations in ways that are strikingly different from those suggested by the standard definition of the core. To appreciate this, consider the following example.[9]

EXAMPLE 13.1. *Suppose that* $N = \{123\}$, *and each coalition is associated with a unique payoff vector:* $\mathbf{u}(\{123\}) = (2,2,2)$, $\mathbf{u}(\{12\}) = (3,3)$, $\mathbf{u}(\{23\}) = (4,1)$, *and* $\mathbf{u}(S) = \mathbf{0}$ *for all other coalitions S. It is easy to see that this game has a unique core state (coalition structure)* $\{\{1\}, \{23\}\}$.

I first describe a deterministic EPCF with unique limit for this example. Because each coalition structure has only one payoff vector, we may equate states with the five coalition structures and schematically write down the PCF. Arrows with coalitional subscripts indicate how the state is being changed, and an absorbing state is shown in boldface:

$x_1 = \{\{123\}\} \rightarrow_{\{2\}} x_4$

$x_2 = \{\{12\}, \{3\}\} \rightarrow_{\{23\}} x_3$

$\mathbf{x_3} = \{\{1\}, \{23\}\} \rightarrow x_3$

$x_4 = \{\{13\}, \{2\}\} \rightarrow_{\{23\}} x_3$

---

[9]Recall that we have used a general way of transforming characteristic functions to F_S-correspondences. In all the examples, we use the particular specification that when a new coalition forms, the induced coalition structure (that immediately results) corresponds to the Γ formulation in Hart and Kurz (1983); see our footnote 6 for a definition.

$$x_5 = \{\{1\}, \{2\}, \{3\}\} \rightarrow_{\{23\}} x_3$$

It is easy to check that if player 2's discount factor exceeds $1/2$, this scheme is indeed an EPCF. Now focus on x_1. This coalition structure is not a core state. The *only* blocking coalition is formed by players 1 and 2. However, if player 1 is farsighted enough, she would not join such a move since she expects that player 2 would "betray" her by forming a move with player 3 to achieve x_3. That is, she would be better off by not deviating from x_1 in the first place. The point is that our EPCF does eliminate the non-core state x_1, *but does not do so by the argument that underlies the definition of the core.*

The reason that x_1 is nevertheless unstable is that player 2 deviates alone, expecting to create a further subsequent move with player 3. Actually, player 2 suffers from a low payoff for one period right after the unilateral deviation, and enjoys higher payoffs for ever from the next period. Thus, player 2's motive for deviating from x_1 is really based on her farsightedness. Thus the *reason* why x_1 is unstable comes from farsightedness, while under the standard definition, x_1 is eliminated for an immediate (myopic) gain. *These are very different arguments, yet they arrive at the same conclusion.*[10]

It should be noted that these propositions are obviously vacuous for games with empty cores. That doesn't mean that a deterministic EPCF won't exist in those cases; see, for instance, Example 13.2 below. Moreover, a general EPCF certainly does exist. Therefore our theory has the feature that it reduces to a well-known solution concept — the core — whenever that solution is nonempty, but continues to make specific predictions otherwise. In this sense, the theory addresses issues of coalition formation in situations that transcend nonempty cores, which is a desideratum that we've already mentioned; see Section 7.1 for a discussion.

---

[10]Notice that under our deterministic EPCF, there may be several profitable moves at a particular state. For instance, it is true that at state x_1, both players 1 and 2 may jointly wish to move if they are given the chance to do so. The reason why 1 also wants to move, in contrast with the argument in the main text, is that if he does not, he foresees disaster coming in the shape of 2 moving anyway, as prescribed by the EPCF. But the point is that the coalition $\{12\}$ is not given the *opportunity* to move. If we do insist on going all the way with this line of reasoning while restraining ourselves to deterministic PCFs, we must allow *only* $\{12\}$ to move — not just today, but tomorrow as well — but as the text argues, this cannot give rise to an EPCF. (To be sure, there may be *stochastic* EPCFs where both coalitions $\{12\}$ and $\{2\}$ obtain the chance to move.) This example therefore also illustrates the conceptual restrictions mentioned at the end of Section 13.3.3.

Finally, there are models of coalition formation which do not come from characteristic functions. The core isn't even well-defined in these cases, but the concept of an EPCF continues to provide new insights, as we shall see below.

13.4.2 Deterministic PCFs and Consistency. We've seen that the tightest restrictions imposed so far — deterministic PCF's with unique absorbing limit — provide an almost-complete characterization of the core. Now I loosen the restrictions slightly by dropping the requirement of a unique limit, though I still require EPCFs to be deterministic and converge from every initial state. That is, I now work with the broader class of *absorbing* deterministic processes of coalition formation.

I first show that this relaxation permits absorbing states that are disjoint from core states, whether or not the core itself is empty.

EXAMPLE 13.2. *Consider the following two characteristic function games. In each game, each coalition has just one feasible payoff vector. Game 1 is as follows:* $N = \{1234\}$, $\mathbf{u}(1234) = (4, 3, 2, 2)$, $\mathbf{u}(234) = (4, 3, 5)$, $\mathbf{u}(\{13\}) = (2, 4)$, $\mathbf{u}(14) = (3, 4)$, $\mathbf{u}(24) = (2, 3)$, $\mathbf{u}(i) = 1$ *for all* $i \in N$, *and* $\mathbf{u}(S) = 0$ *for all other coalitions S.*

Game 2 retains all the features of Game 1, but changes $\mathbf{u}(234)$ *to* $(4, 3, 4)$ *and* $\mathbf{u}(14)$ *to* $(3, 5)$. *Game 1 does not have any core state (weak or strong), and Game 2 has a unique (weak and strong) core allocation, given by* $\{\{14\}, \{2\}, \{3\}\}$.

Assume that all players have a common discount factor δ.

Because each coalition structure has only one payoff vector, there are fifteen states in each of these two games. Once again, we describe an absorbing deterministic PCF in a schematic way, just as in Example 13.1:

$x_1 = \{\{\mathbf{1234}\}\} \rightarrow x_1$

$x_2 = \{\{123\}, \{4\}\} \rightarrow_{\{1234\}} x_1$

$x_3 = \{\{124\}, \{3\}\} \rightarrow_{\{1234\}} x_1$

$x_4 = \{\{134\}, \{2\}\} \rightarrow_{\{24\}} x_8$

$x_5 = \{\{234\}, \{1\}\} \rightarrow_{\{13\}} x_8$

$x_6 = \{\{12\}, \{34\}\} \rightarrow_{\{1234\}} x_1$

$x_7 = \{\{12\}, \{3\}, \{4\}\} \rightarrow_{\{13\}} x_9 \rightarrow_{\{24\}} x_8$

$\mathbf{x_8} = \{\{\mathbf{13}\}, \{\mathbf{24}\}\} \rightarrow \mathbf{x_8}$

$x_9 = \{\{13\}, \{2\}, \{4\}\} \rightarrow_{\{24\}} x_8$

$x_{10} = \{\{14\}, \{23\}\} \rightarrow_{\{1\}} x_{12} \rightarrow_{\{1234\}} x_1$

$x_{11} = \{\{14\}, \{2\}, \{3\}\} \rightarrow_{\{1\}} x_{15} \rightarrow_{\{1234\}} x_1$

$x_{12} = \{\{23\}, \{1\}, \{4\}\} \rightarrow_{\{1234\}} x_1$

$x_{13} = \{\{24\}, \{1\}, \{3\}\} \rightarrow_{\{13\}} x_8$

$x_{14} = \{\{34\}, \{1\}, \{2\}\} \rightarrow_{\{1234\}} x_1$

$x_{15} = \{\{1\}, \{2\}, \{3\}, \{4\}\} \rightarrow_{\{1234\}} x_1.$

It is very easy to verify that if $\delta \geq \frac{3}{4}$, then this describes an equilibrium PCF for both games 1 and 2. (I omit the tedious details.)

The absorbing states under the EPCF are x_1 and x_8. Neither of these is a core allocation. The state x_1 involves the grand coalition, which is blocked by the coalition $\{234\}$. But that block is counterblocked, in turn, by the formation of the coalition $\{13\}$, which takes us to the absorbing state x_8. If player 2 is farsighted, she will know the full consequences of her initial deviation will generate a long-term payoff of only 2, and so she will refuse to participate in the blocking of x_1.

In similar vein, x_8, which features the structure $\{\{13\}, \{24\}\}$, is blocked by $\{14\}$ — with associated structure $\{\{14\}, \{2\}, \{3\}\}$ (the state x_{11}). But matters don't end there: our EPCF will transit thereafter to the state x_{15} and from there to x_1. A farsighted player 4 won't want to start this chain of deviations.

Put another way, players 1 and 2 prefer x_1 to x_8, while players 3 and 4 prefer x_8 to x_1. However, starting from x_1, players 3 and/or 4 can move only to x_2, x_3, and x_6. All of these states will come back to x_1. Thus, players 3 and 4 cannot move the state to x_8 without the help of players 1 and/or 2. A parallel argument applies to players 1 and 2 at x_8.

We have therefore shown that there may be an absorbing deterministic EPCF with no core elements among its absorbing states.

Indeed, this is true whether the core is empty or not. To verify this, consider the game 2 variant. It has a unique core allocation, which is the state x_{11}. Yet x_{11} is destabilized under our EPCF by the departure of player 1, inducing the coalition structure of singletons x_{15}. Player 1 wouldn't do this if matters were to end here, but they don't. Instead, she correctly anticipates that the system will move thereafter to the absorbing state x_1, in which she *is* better off relative to the core outcome!

These arguments do not contradict our earlier propositions on core equivalence. The reason they don't is that they rely on the fact that the EPCF under discussion *has multiple absorbing states*. Of course, we know from Proposition 13.2 that in game 2 of the example, there is *another* EPCF that uniquely picks out the core state. But there is no obvious selection criterion that permits us to focus on that equilibrium as opposed to the one we've highlighted here.

So the core does possess a remarkable consistency property: a deterministic EPCF with a *unique* limit picks out a core point. But when the self-referential nature of the process is dropped (by admitting more than one absorbing state), then the possibilities widen beyond the core.

It turns out, however, that all absorbing deterministic EPCF's have absorbing states that lie within the *largest consistent set*, a concept due to Chwe (1994). We've already encountered the sequential blocking relation that lies at the heart of this solution concept; see Section 12.2.1 from the previous chapter. We recall that notion now and apply it to the state-based model at hand.

Consider any model of coalition formation (not just a characteristic function). Say that a state y *sequentially dominates* some other state x, if there exists a sequence of states $\{x^0, x^1, \dots, x^m\}$ in X with $x^0 = x$ and $x^m = y$, and coalitions $S_0, S_1, \dots, S_{m-1}$ such that for $j = 0, \dots, m - 1$, $x^{j+1} \in F_{S_j}(x^j)$ and $u_i(y) \geq u_i(x^j)$ for all $i \in S_j$.

This is just the sequential blocking idea in Section 12.2.1, except that we ask only that all players be *weakly* better off at the "final" state y. Chwe's definition requires that they all be strictly better off. This is not a major difference: in the definition to follow we will simply use indifferences in either direction, both as a rationale for a player to deviate or to remain where she currently is.

Say that a collection Y of states is *consistent* if the following holds: $x \in Y$ if and only if for every coalition S and for any state $z \in F_S(x)$, there exists $y \in Y$, where either $y = z$ or y sequentially dominates z, such that the inequality $u_i(x) \geq u_i(y)$ holds for at least one player $i \in S$.

In words, a collection of states is consistent if every coalitional move from any element of that collection leads to a "domination chain" (starting with the move and ending within that same collection of states) such that at the "end" of that chain, there is some member of the original deviating coalition who feels that the move was not worthwhile.

Proposition 1 in Chwe (1994) establishes that there is a *largest* consistent set among all consistent sets: a set which is itself consistent and which contains every consistent set.[11]

I now link — at least in one direction — Chwe's largest consistent set to the limit states of absorbing deterministic EPCFs.

PROPOSITION 13.4. *There exists $\delta^* \in (0, 1)$ such that if $\delta_i \in (\delta^*, 1)$ for all i, then the set of all absorbing states of any absorbing deterministic EPCF is contained within the largest consistent set.*

We postpone a formal proof of this result to Section 13.6.

Notice that this proposition is true of any game of coalition formation, and not just characteristic functions. It is also worth noting that the largest consistent set may be "large" but is certainly not exhaustive. For instance, in the Prisoners' Dilemma — transformed into a dynamic model of coalition formation along the lines discussed in Section 13.3.2 — the largest consistent set is a singleton consisting of the cooperative outcome.

Nevertheless, there are reasons to believe that the largest consistent set may be too inclusive — "too large" — in some situations (see, for example, the discussion in Xue (1998)). Our notion of an EPCF brings out one reason for this, as the following example illustrates:

EXAMPLE 13.3. $N = \{123\}$. *There are four states, fully described by the payoff vectors they generate: $x_1 = (2, 2, 2)$, $x_2 = (0, 0, 0)$, $x_3 = (6, 6, 0)$ and $x_4 = (1, 0, 6)$. Describe F_S as follows: at x_1, coalition $\{12\}$ or player 3 are the only coalitions that can move, and the only move (in either case) is to*

---

[11]As noted earlier, I use the weak domination ordering, but Chwe's proposition extends to this case with no changes.

x_2. *At x_2, only coalitions {2} and {13} can move, and in either case, they can induce either x_3 or x_4. From no other state is any move possible, and no other coalition is capable of any other move.*

It is easy to see that the largest consistent set consists of the three states (x_1, x_3, x_4). In particular, the state x_1 is a member of this set for the following reason: the coalition {12} avoids inducing x_2 because it anticipates the continuation by {13} to x_4, and player 3 similarly negates a move to x_2 because she fears the subsequent creation of state x_3 (by player 2).

However, there is no deterministic absorbing EPCF — and indeed, *no* EPCF at all — with x_1 as an absorbing state (provided that discount factors are close enough to unity). To see this, let p be the probability that some EPCF assigns to {2} moving at x_2 (so that $1 - p$ is assigned to {13}). Neglecting discounting for a moment, note that if $p > 1/3$, then {12} will want to move from x_1, whereas if $p < 2/3$ player 3 will want to move from x_1. It is now trivial to see that for all discount factors close enough to 1, x_1 cannot be an absorbing state.

This example shows quite starkly why the notion of consistency is less restrictive than that of an EPCF. Two domination chains along two sequential blocks may have different moves attached to them starting from the same state. Thus, as seen above, in the largest consistent set, coalition {12} entertains one sort of conjecture about what will happen at x_2 and player 3 entertains another. If all players have common beliefs (as they must in an EPCF), then this possibility cannot arise. This is one reason why the set of all absorbing states (under all deterministic absorbing EPCFs) is typically a *strict* subset of the largest consistent set. Another reason for strict inclusion has to do with the efficiency of coalitional moves.[12]

13.4.3 Deterministic Schemes: Absorption, Cycles and Efficiency.
Example 13.3 in the previous section shows that the set of absorbing states (under deterministic absorbing EPCFs) can be a strict subset of the largest consistent set. In particular, our EPCF prunes inefficient outcomes from that set. This suggests that our dynamic structure may be generally adept at eliminating inefficient outcomes. By virtue of Proposition 13.3, this is certainly

---

[12]See Xue (1998), in which similar issues are raised in the context of a static model of coalitional moves.

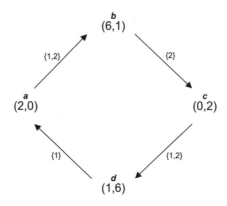

FIGURE 13.1. PAYOFFS AND APPROVAL COMMITTEES FOR EXAMPLE 13.4

true of absorbing schemes that have unique limit, provided that all players are sufficiently farsighted and the underlying game is representable by a characteristic function. However, when an EPCF exhibits multiple absorbing states, one or more of them may well be inefficient.

EXAMPLE 13.4. $N = \{12\}$ and $X = \{a, b, c, d\}$. *Payoffs and approval committees are as described in Figure 13.1.*

In the class of absorbing deterministic schemes there is exactly one equilibrium, provided the discount factor of each player exceeds 2/3. This equilibrium has absorbing states $\{a, c\}$. Notice that the payoffs from these states are inefficient.[13]

To see why, first note that in any absorbing deterministic equilibrium, neither b nor d can be absorbing states. For suppose, on the contrary, that b is absorbing. Then notice that a cannot be absorbing; indeed, that $a \to b$. This means that a move from b to c, engineered by player 2, has the following possible payoff continuations for player 2: the constant payoff $(2, 2, \ldots)$, the path $(2, 6, 6, \ldots)$, and the path $(2, 6, 0, 1, 1 \ldots)$. In *each* of these cases 2 earns a (normalized) discounted payoff that exceeds 1, which contradicts the presumption that b is absorbing. The argument that d cannot be absorbing is an exact parallel.

So a or c (or both) must be absorbing, and there are no other absorbing states.

[13]$\{a, c\}$ is also the largest consistent set.

Next, observe that *both* a and c must simultaneously be absorbing. For say only a were absorbing. Then it must be the case that $d \rightarrow a$. Now consider a move from a to b, engineered by {12}. It is obvious that player 2 must earn positive payoff from this move. Moreover, for any $\delta_1 < 1$, player 1's (normalized) discounted payoff must strictly exceed 2 (we use here the observation that $d \rightarrow a$). This contradicts the assumption that *only a* is absorbing. A parallel argument holds for c. Therefore both a and c are absorbing.

This leaves us with only one possible absorbing deterministic EPCF, in which $d \rightarrow a$ and $b \rightarrow c$. Indeed, such a PCF *is* an equilibrium, provided that the discount factor of each player exceeds 2/3.

Readers interested in understanding better the source of this efficiency failure are referred to Appendix A, item 5.

To conclude this section, consider the following PCF which, while deterministic, has no absorbing states: $a \rightarrow b \rightarrow c \rightarrow d \rightarrow a$. Provided that discount factors are close to unity, it is easy to check that each move prescribed by the scheme is strictly profitable (and efficient in the class of all profitable moves). Therefore this cyclical scheme is an EPCF. For discount factors close to one, the (normalized) discounted payoff to each player is approximately 2.25. This payoff vector is efficient. In what follows, we move on to a closer investigation of cyclical and stochastic schemes.

13.5 Stochastic Equilibrium Processes

In the remainder of this chapter, we concentrate on stochastic processes of coalition formation. Uncertainty enters the story in two distinct ways. First, at any stage, several coalitions may have profitable moves. *Which* coalition gets to move may well be probabilistically chosen. Second, it is possible that a particular coalition has more than one efficient move, and that it might randomize among them. The discussion that follows shows that in many typical situations one or more of these randomizations may be necessary in order to generate an equilibrium.

The uncertain nature of the process may or may not be intertwined with cycles — possibly stochastic reversions of the state of the game to some given state. Formally, a (nonsingleton) collection of states $(x^1, \ldots, x^k)$ under a PCF forms a (stochastic) *cycle* if $p(x^i, x^{i+1}) > 0$ for all $1 \leq i < k$ and $p(x^k, x^1) > 0$. A PCF that exhibits a cycle will be called *cyclical*.

The purpose of the analysis that follows is to understand these phenomena, largely through the use of examples. A large part of our discussion will take place explicitly in the context of strategic form games.

13.5.1 Randomization and Cycles: An Example . The purpose of this section is to illustrate the need for cycles and/or randomization in certain situations. We do this by considering the following restatement of the "roommate problem". This is a situation with three players, any of two of whom can "share a room". In each case, the player left out obtains zero. Moreover, for each pair of roommates, there is one who obtains a payoff of 1, while the other obtains a payoff of a (to be parametrically varied in the example). Details follow, couched in the language of a model of coalition formation.

EXAMPLE 13.5. *Let* $N = \{123\}$, $X = \{x, y, z\}$, $F_{\{23\}}(x) = \{x, y\}$, $F_{\{13\}}(y) = \{y, z\}$, $F_{\{12\}}(z) = \{z, x\}$, *and* $F_S(x') = \{x'\}$ *for all other combinations of* x' *and* S. *Players have a common discount factor* δ. *Payoffs for each state* x' *are described in the following array:*

	x	y	z
1	1	0	a
2	a	1	0
3	0	a	1

Note that it is easy to rewrite this example in the more familiar guise of a characteristic function.[14] Appendix B contains a demonstration of the following

OBSERVATION 13.1. *For each* $a > 0$ *and* $\delta \in (0, 1)$, *the game in Example 13.5 admits a symmetric EPCF* p. *For* $a \leq \frac{1}{1+\delta}$, $p(x, y) = p(y, z) = p(z, x) = 1$, *and for* $a > \frac{1}{1+\delta}$, $p(x, y) = p(y, z) = p(z, x) = \frac{(1-a)(1-\delta)}{\delta(2a-1)}$.

Moreover, for $a > \frac{1}{1+\delta}$, *no deterministic EPCF exists.*

This example (and the accompanying Observation) is designed to illustrate several points.

---

[14]Let $N = \{123\}$ with $\mathbf{U}(\{123\}) = \{(0, 0, 0)\}$, $\mathbf{U}(\{12\}) = \{(1, a)\}$, $\mathbf{U}(\{23\}) = \{(1, a)\}$, $\mathbf{U}(\{3, 1\}) = \{(1, a)\}$, and $\mathbf{U}(\{i\}) = \{0\}$ for any $i \in N$.

First, there is no hope of a general existence result for deterministic schemes, even if we restrict ourselves to characteristic functions. This is also true of strategic form games, though we will see situations in which stochastic EPCFs coexist with their deterministic counterparts.

Second, observe that once cycles and randomization make an appearance, the cardinality of payoffs matters in the determination of a particular equilibrium. In the example, if a is small enough (that is, if $a \leq \frac{1}{1+\delta}$), then the equilibrium cycle is deterministic, and there is no chance of remaining in the same state in any period. However, as a goes up from $a = \frac{1}{1+\delta}$, the probability of moving to the next state comes down. The cycles become stochastic. As a increases further, each coalition structure becomes relatively more stable as the cyclical movement becomes slower in the stochastic sense, although no state ever becomes absorbing.

The consequences of an increase in the discount factor are similar. provided that $a > \frac{1}{1+\delta}$, the cycles slow down as δ increases. If δ is very close to unity, the cross-state transitions measured by p become very close to zero. Notice that p never *becomes* zero — the tension of a possible move is needed to sustain the scheme.[15] Nevertheless, we still can say that if δ goes to unity, each coalition structure becomes more stable in a stochastic sense.

Third, the EPCF in the example illustrates only one of the two sources of stochastic behavior discussed earlier. At each state, there is exactly one potential deviating coalition. Yet an EPCF can (and sometimes, as in the example), *must* be stochastic. Randomization occurs not over multiple deviating coalitions, but over what a single coalition actually does; "moving" versus "staying" in this example. This type of randomization can occur only when at least one member of the deviating coalition is indifferent between moving and staying.[16] In the next section, we will see several examples of the second source of uncertainty: the sort that stems from randomization over multiple coalitional deviations.

---

[15]However, asymmetric roommate problems (in which the cardinalities of vNM utility functions or the values of discount factors differ across agents) may have well absorbing states.

[16]Often, this makes it easy to compute equilibrium payoffs. As long as there is only one possible coalitional deviation at each state, and that coalitional move is randomized, an indifferent player's normalized payoff must be exactly the same as the atemporal payoff from the current state.

13.5.2 Temporary Agreements for Games in Strategic Form.
In this section, we apply our method to strategic form games. We use the formulation described in item (3) of Section 13.3.2. A state is simply an action profile **a** of the strategic form "stage" game. A coalition S has the power to unilaterally alter $\mathbf{a}_S$ in any period in which it is active, while the remaining players are "frozen" at $\mathbf{a}_{-S}$.

This is not the formulation that we have adopted elsewhere in the book. The very device of a partition function presumes that even when some coalition has a chance to form or regroup, the remaining coalitions can — at the very least — change their actions. By presuming, as we do here, that players outside the active coalition do not change their actions, we are looking at a very different model.

One objective of our analysis is to study how the possibly stochastic nature of coalition formation affects efficiency in strategic-form games.

13.5.2.1 Games with Common Payoffs.
It will be useful to begin with a situation in which efficiency is *not* impaired, and this will serve as a benchmark for the more interesting cases to follow. To this end, consider the class of all strategic games with *common payoffs*, which yield similar payoffs to all players for any action profile. To be sure, such games are not without genuine strategic significance; for instance, the following well-known pure coordination game (with a and b negative) is a special case of what I have in mind:

	L	R
T	$1,1$	b,b
B	a,a	$0,0$

Formally, consider a game in strategic form. N is a set of players. Player i has action set A_i. **A** denotes the product of all action sets. Player i has a payoff function u_i defined on A. This game is one of *common payoffs* if $u_i(\mathbf{a}) = u_j(\mathbf{a})$ for every action profile $\mathbf{a} \in A$.

It is easy to embed this game into an intertemporal model of coalition formation. A state is just the ongoing action profile **a**, and $F_S(\mathbf{a})$ is the set of all states $\mathbf{a}'$ such that $\mathbf{a}'_S \in \times_{i \in S} A_i$, and $a'_i = a_i$ for all $i \notin S$.

In words: an action vector is available to S if it is feasible for its members and if the remaining players leave their actions unchanged.

To complete the description, assume each player i has a common discount factor $\delta \in (0, 1)$.

PROPOSITION 13.5. *Consider a game of common payoffs with the property that there is a unique action profile* **a*** *at which all player payoffs are maximized. Then every EPCF for that game with a common discount factor involves* $p(a^*, a^*) = 1$ *and has* a^* *as the unique absorbing limit starting from any* $a \in A$.

See Section 13.6 for the proof of this proposition.

That at least one such EPCF exists with the claimed property is trivially true. That EPCF would correspond, for instance, to the playing of the "good equilibrium" in a coordination game. The extra bite of this result lies in the assertion of asymptotic optimality for *every* EPCF.

There is a related literature which seeks to eliminate "bad equilibria" in coordination games. For instance, Lagunoff and Matsui (1997) contains a related result (see also Corollary 2 in Kandori, Mailath and Rob (1993)). Lagunoff and Matsui study repeated pure coordination games in which only one player can change her action in each period, and show that if the discount factor is close to unity there is a unique subgame perfect equilibrium in which the action profile converges to the Pareto-efficient one. To be sure, there are important differences, not the least of which is that our approach permits the writing of temporarily *binding* agreements.

Now for the bad news. Binding agreements notwithstanding, the finding of ubiquitous cooperation in common-payoff situations does not extend even to *coordination* games with non-common payoffs. The following example describes a 2×2 game in which there is an EPCF with an inefficient absorbing limit.

EXAMPLE 13.6. *Consider the following 2×2 strategic form game, with common discount factor δ.*

	L	R
T	1, 1	−5, −1
B	−1, −5	0, 0

Then the induced game of coalition formation has an EPCF with unique absorbing limit (B, R) *whenever* $\delta \geq \frac{2}{3}$.

Write the states (T,L), (B,L), (T,R), (B,R) as x, y, z, and w, respectively. Consider the PCF described by $p(x,y) = p(x,z) = \frac{1}{2}$, $p(y,w) = p(z,w) = p(w,w) = 1$. Then

$$V_1(x,p) = V_2(x,p) = 1 + \delta\left(\frac{-1-5}{2}\right) = 1 - 3\delta,$$
$$V_1(y,p) = V_2(z,p) = -1,$$
$$V_1(z,p) = V_2(y,p) = -5,$$
$$V_1(w,p) = V_2(w,z) = 0.$$

As we can easily see from these expressions, there is an incentive for either player to move from x as long as $1 - 3\delta \leq -1$, which is equivalent to $\delta \geq \frac{2}{3}$. So the PCF is indeed an EPCF under this condition.

The striking feature of this EPCF is that although x is the highest payoff state for every player, it is not stable. The temporary agreement x is upset by unilateral deviations, in which each deviation is bolstered by the fear of the other player's deviation. Observe that this sort of "meta-coordination failure" relies intimately on the failure of common payoffs. It is *not* the "standard" coordination failure that would be present even in games with common payoffs.[17]

Finally, recall that we mentioned two sources of stochastic equilibria; one stemming from the multiplicity of coalitional moves by a single deviating coalition and the other from the multiplicity of deviating coalitions. The EPCF in Example 13.6 represents an instance of the second type of uncertainty and its effects. Randomization among profitably deviating coalitions may cause inefficiency in the resulting outcome.[18]

13.5.2.2 The Prisoners' Dilemma. The Prisoners' Dilemma is a leading example in game theory. I therefore study the EPCFs of this game in some detail.

---

[17]Equilibrium selection in Kandori, Mailath and Rob (1993) or Young (1993) is related to the risk-dominance of an action profile (Harsanyi and Selten (1988)), and indeed, something similar plays an important role in the example. The Pareto superior Nash equilibrium is a risk dominated equilibrium $(1 - (-1) < 0 - (-5))$. However, in general, the conditions for the existence of an EPCF with breakdown in cooperation are different even in coordination games.

[18]Of course, there are other EPCFs: for instance, a "cooperative" EPCF with $p(x,x) = p(y,x) = p(z,x) = p(w,x) = 1$ exists for any value of δ. What Example 13.6 says is that there can be another EPCF that attains a Pareto inferior state as the unique absorbing state even in a coordination game unless we have common payoffs.

Consider the following 2×2 strategic form game:

	L	R
T	1, 1	b, a
B	a, b	0, 0

where $a > 1$ and $b < 0$. As in Example 13.6, write (T, L), (B, L), (T, R), (B, R) as x, y, z, and w respectively.

Unlike games with common payoffs, x no longer attains the highest possible payoff, and it is well-known that w is the unique dominant strategy Nash equilibrium of this game. Our model of coalition formation yields a more varied set of results, which we now attempt to describe. We begin with an observation for deterministic EPCFs:

OBSERVATION 13.2. *The Prisoners' Dilemma admits various deterministic EPCFs depending on specific parameter values:*

a. *There is a deterministic EPCF with its unique absorbing limit at x ($p(x, x) = p(y, w) = p(z, w) = p(w, x) = 1$) if and only if $a \leq 1 + \delta$.*

b. *There is a deterministic EPCF with its unique limit at w ($p(x, y) = p(y, w) = p(z, w) = p(w, w) = 1$) if and only if $a \geq \frac{1}{1-\delta}$ and $b \leq -\frac{1}{\delta}$.*

c. *There is a deterministic cyclical EPCF ($p(x, y) = p(y, w) = p(z, w) = p(w, x) = 1$) if and only if $a \geq 1 + \delta$ and $b \geq -\frac{1}{\delta}$.*

Notice that the observation provides a *complete* characterization of all deterministic EPCFs. Case (a) permits cooperation to be sustained as the unique limit of a deterministic EPCF as long as (and *only if*) the "defection payoff" a is not too large. Although this finding is not unintuitive, it provides a different perspective on the relationship between our solution concept and the largest consistent set (see above, Proposition 13.4). It is easy to see that the largest consistent set is simply the singleton $\{x\}$ no matter what values a and b take (provided, of course, that $a > 1$ and $b < 0$). However, no EPCF supports x if the defection payoff is too large *even when* δ is close to unity.

This observation does not contradict Proposition 13.4, in which a *deterministic* EPCF with absorbing limit was shown to lie within the LCS. The point is that once the defection payoff is large enough, such EPCFs fail to exist. Cycles occur (as Case (c) illustrates), but Proposition 13.4 is silent on cyclical EPCFs. Another seeming contradiction to Proposition 13.4 is Case (b), which asserts that a deterministic EPCF may support w as a unique absorbing limit. Notice, however, that for every given value of the defection payoff, the existence of such a scheme is conditional on δ not being too large, whereas Proposition 13.4 only applies for discount factors sufficiently close to unity. For given values of the other parameters, Case (b) must disappear as the discount factor approaches unity.

Observation 13.2 has the following implications. First, deterministic equilibria do not exist over the full spectrum of parameters, even as the discount factor goes to 1. Therefore stochastic EPCFS *must* be invoked for a full description of outcomes in the Prisoners' Dilemma. Second, provided that the discount factor is close enough to one *and* provided a deterministic EPCF with unique absorbing limit exists, then it can only sustain cooperation rather than defection. At the same time, if the defection payoff is too large (larger than 2, to be precise) then the existence of such EPCFs is jeopardized: Case (c) shows that in such cases one typically cycles between cooperation and defection. Moreover, these cycles run through the state w so they cannot be efficient.

Inefficiency is also a feature of stochastic EPCFs. This is not to say that stochastic equilibria cannot be asymptotically efficient. They can. But they can be inefficient in a way that deterministic equilibria with absorbing limits cannot be. The next Observation clarifies this point, and also fills in the "existence gap" left by deterministic EPCFs:

OBSERVATION 13.3. *Here are some instances of stochastic EPCFs in the Prisoners' Dilemma:*

a. *There is a stochastic, symmetric, EPCF of the form $p(x, y) = p(x, z) = \frac{1}{2}$, $p(y, w) = p(z, w) = p(w, w) = 1$ if and only if $b \leq -a - \frac{2}{\delta}$. This EPCF has a unique absorbing limit at noncooperation.*

b. *There is a stochastic cyclical EPCF with $p(x, y) = p(x, z) = p$ and $p(y, w) = p(z, w) = p(w, x) = 1$ if and only if*

$$-a - \frac{2}{\delta} \leq b \leq \left(\frac{1 + \delta + \delta^2}{\delta + \delta^2} \right) a - \frac{2}{\delta}.$$

Moreover, if $a \geq 1 + \delta$ (resp. $a < 1 + \delta$), then $p = 1/2$ (resp. $p < 1/2$).

I recognize that there may be other stochastic EPCFs, but I choose to focus on the ones in Observation 13.3 for two reasons. First, the five classes of EPCFs described in the two observations do span

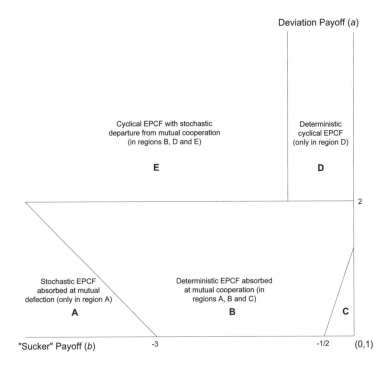

FIGURE 13.2. SOME SYMMETRIC EPCFs IN THE PRISONERS' DILEMMA

the entire space of parameters, so we can see at least one kind of EPCF for each configuration of parameters. Figure 13.2 illustrates this very clearly for the limiting case as discounting vanishes. As already discussed, the class (b) in Observation 13.2 vanishes in this limiting case, but the remaining four categories of equilibrium do span the entire domain.

Indeed, in the class of symmetric EPCFs, all we've left out is the stochastic EPCF which converges to the cooperative outcome. It can be checked that this looks very similar to its deterministic counterpart, and exists under similar conditions on the parameters.

Second, it is clear that stochastic EPCFs tell a different story, at least as far as absorption into an inefficient limit is concerned. Case (a) in Observation 13.3 tells us that in contrast to the deterministic case, it *is* possible (even when $\delta \simeq 1$) to construct stochastic schemes with unique absorbing limit at mutual defection. The condition for this to happen can be interpreted in the form of a low enough value of b, the so-called "sucker payoff". That b matters is not surprising, as this payoff (induced by the other agent's departure

from cooperation) is what creates the "meta-coordination failure" discussed in the context of Example 13.6.

Finally, Case (b) identifies (necessary and sufficient) conditions for the presence of stochastic EPCFs that exhibit cycles. Notice that if a is not too large then the cooperative outcome must exhibit some inertia along this cycle ($p(x, x) > 0$).

It may be worth pointing out that the conditions identified in Case (c) of Observation 13.2 and Case (b) of Observation 13.3 apply for the entire range of values for a and b. In particular, we can use these conditions to conclude that no (symmetric) coordination game can exhibit a cycle.

The preceding discussion also makes clear that cardinalities do matter in determining the sort of EPCFs that characterize any given Prisoners' Dilemma. To emphasize this and to focus on the leading case in which δ is close to unity, we end this section (and the chapter) with three examples.

EXAMPLE 13.7. (Prisoners' Dilemma 1.) *No EPCF supports the unique dominant strategy Nash equilibrium as its absorbing state, but cooperation can be supported as the unique absorbing state of a deterministic EPCF for δ close to unity.*[19]

	L	R
T	1, 1	$-\frac{1}{2}, \frac{3}{2}$
B	$\frac{3}{2}, -\frac{1}{2}$	0, 0

EXAMPLE 13.8. (Prisoners' Dilemma 2.) *The Nash equilibrium can be supported as the unique absorbing state of a stochastic EPCF for δ close to unity.*

	L	R
T	1, 1	−6,3
B	3, −6	0, 0

EXAMPLE 13.9. (Prisoners' Dilemma 3.) *Both cooperation and noncooperation states may be supported as the unique absorbing state of EPCFs (deterministic and stochastic, respectively) for δ close to unity.*

---

[19]This game also has a stochastic cyclical EPCF with $p(x, x) \in (0, 1)$.

	L	R
T	1, 1	$-4, \frac{3}{2}$
B	$\frac{3}{2}, -4$	0, 0

13.6 Proofs

Proof of Proposition 13.2. We will construct a deterministic PCF p that absorbs into x^* no matter where it starts from. Write out x^* explicitly as $(\pi^*, \mathbf{u}^*)$, where π^* is the coalition structure $\{S_1^*, \dots S_K^*\}$. Let $x = (\pi, \mathbf{u})$ be any initial state.

CASE 1. $x = x^*$. Set $p(x, x) = 1$.

CASE 2. Case 1 does not hold, and there exists a player i such that $\{i\} \notin \pi$, and $u_i^* > u_i$. Pick the smallest index i with this property, and set $p(x, y) = 1$, where $y = (\pi', \mathbf{u}') \in F_i(x)$ with $u_i' = \max \mathbf{U}(\{i\})$.

CASE 3. Cases 1 and 2 do not hold, and there exists a coalition $S \in \pi^*$ such that $u_i^* > u_i$ for all $i \in S$. Pick the smallest index k such that S_k^* has this property, and set $p(x, y) = 1$, where $y = (\pi', \mathbf{u}') \in F_{S_k^*}(x)$ with $\mathbf{u}'_{S_k^*} = \mathbf{u}^*_{S_k^*}$.

CASE 4. Cases 1, 2 and 3 do not hold, and there exists a coalition $S \in \pi^*$ such that $u_i^* \geq u_i$ for all $i \in S$, and *either* $S \notin \pi$, or $S \in \pi$ and $\mathbf{u}_S^* \neq \mathbf{u}_S$. Pick the smallest index k such that S_k^* has this property, and set $p(x, y) = 1$, where $y = (\pi', \mathbf{u}') \in F_{S_k^*}(x)$ with $\mathbf{u}'_{S_k^*} = \mathbf{u}^*_{S_k^*}$.

For this construction to be sensible, at least one of the situations described must obtain. To see this, assume that Cases 1–3 do not hold. We show that Case 4 must hold. To this end, pick any coalition T in π. If it is a singleton, we must have $u_i \leq u_i^*$ (because a^* is a core allocation). We claim the same is true of all $i \in T$ even if T is not a singleton.

For if this is false, then $u_j > u_j^*$ for some $j \in T$. But then, because $\mathbf{u}^*$ is a strong core allocation, there exists $i \in T$ such that $u_i < u_i^*$. Clearly $\{i\} \notin \pi$ (because $T \in \pi$ and T is not a singleton). But this means that Case 2 holds, a contradiction.

So we have shown that $\mathbf{u}^* \geq \mathbf{u}$. In particular, for any $S \in \pi^*$, we have $\mathbf{u}_S^* \geq \mathbf{u}_S$. To complete the argument, suppose that for all $S \in \pi$, we have $\mathbf{u}_S^* = \mathbf{u}_S$. Then π *cannot* equal π^* (otherwise we would be in

Case 1). This means that there must exist $S \in \pi^*$ (with $\mathbf{u}_S^* \geq \mathbf{u}_S$, as already shown) such that $S \notin \pi$. So Case 4 holds whenever Cases 1–3 do not.

Therefore the (deterministic) transition from x to y is well-defined in all cases. It is also easy to see that apart from x^*, $x \neq y$ for every other state, and that there are no cycles. It follows that x^* is the unique absorbing limit of this PCF.

To complete the proof, we must show that all the conditions of an EPCF are satisfied by this PCF. To assure this, we first choose the threshold value of δ^*. For any individual i, let m_i be the maximal payoff that he enjoys over all states in which he receives less than his core payoff u_i^*. Define δ_i^* by $\delta_i^{*C} u_i^* = m_i$, where C is the total number of states, and $\delta^* \equiv \max_{i \in N} \delta_i^*$. We take it that the discount factor of every player strictly exceeds this threshold.

Begin with the state x^*, and consider any move by any coalition S to $x = (\pi, \mathbf{u})$. Let L be the members of S who are no better off in the "static sense" by doing so: $L = \{j \in S | u_j \leq u_j^*\}$. Observe that L is nonempty. Now apply our constructed PCF thereafter. Notice that the payoff to any member of S can only change if some member of L initiates a future move (and indeed, this must happen under the PCF). Let $i \in L$ be one of the first movers from S after the initial move by S. Given the PCF, i cannot enjoy any more than his core payoff u_i^* after this move is made. The same is also true for the intervening period between the first move by S and the later move by i. We may conclude that i *cannot* be strictly better off (relative to the core payoff) by participating in the move by S. It follows that at x^*, no strictly profitable move exists, so we are justified in placing $p(x^*, x^*) = 1$.

Now consider some state $x \neq x^*$. Suppose that we are in case 2. Given the definition of our PCF, it only needs to be shown that the stipulated move is profitable. Notice that

$$V_i(x, p) = (1 - \delta_i)u_i + \delta_i V_i(y, p)$$

while

$$V_i(y, p) \geq \delta_i^C u_i^* > \delta^{*C} u_i^*$$

by the normalization that all payoffs are nonnegative, the fact that the strong core allocation is reached under the PCF in at most C periods, and the fact that $\delta_i > \delta^*$ for all i. Combining these last two expressions, it is easy to see that

$$V_i(y, p) - V_i(x, p) > (1 - \delta_i)\left[\delta^{*C} u_i^* - u_i\right] \geq (1 - \delta_i)[m_i - u_i] \geq 0,$$

where the very last inequality follows from the fact that $u_i^* > u_i$, and the definition of m_i.

Now suppose that we are in cases 3 or 4. Then there is some coalition S_k^* which is required to move directly to its segment of the strong core allocation, creating the state y. Moreover, by condition (c) in the definition of the move correspondence F_S, and given our PCF, S_k^* will receive this payoff for ever after. Clearly this move is weakly profitable. To see that it is efficient, consider any other state $z = (\pi', \mathbf{u}') \in F_{S_k^*}(x)$. Following the same line of reasoning as in case 1, let L be the subset of people in S_k^* who are no better off (relative to their core payoff) by doing so: $L = \{j \in S | u_j \le u_j^*\}$. Observe that L is nonempty. Now follow a parallel argument to see that there exists $i \in S_k^*$ who cannot derive a higher payoff from the route precipitated by this alternative move by S_k^*. In other words, the prescribed move for S_k^* is efficient.

Finally, notice that our ordering of the cases guarantees that some strictly profitable (and efficient) move is always made whenever one exists. □

Proof of Proposition 13.3. Our first task is to fix δ^*. For any $x = (\pi, \mathbf{u})$ that is not a weak core state, there is some coalition S and $\mathbf{u}' \in U(S)$ such that $\mathbf{u}' \gg \mathbf{u}_S$. Pick $\epsilon > 0$ such that $\mathbf{u}' \ge \epsilon + \mathbf{u}_S$. Because there are only a finite number of states, we may choose ϵ so that this inequality holds uniformly across all noncore states, all coalitions S, and all feasible payoffs $\mathbf{u}' \in U(S)$ that do better for S. Next, denote by M the maximal (one-period) payoff accruing to any player under the characteristic function. Finally, define δ^* so that $(1 - \delta^{*C})M < \epsilon$, where C is the total number of states.

Consider any associated intertemporal model of coalition formation, with $\delta_i > \delta^*$ for all $i \in N$. Suppose, contrary to the statement of the theorem, that there exists a deterministic EPCF p with unique limit x, where x is *not* a weak core state. Then there is some coalition S and $\mathbf{u}' \in U(S)$ such that $\mathbf{u}' \gg \mathbf{u}_S$. Let S induce the state $y = (\pi'', \mathbf{u}'') \in F_S(x)$ such that $\mathbf{u}_S'' = \mathbf{u}'$. Given the EPCF, starting from y, the system must attain x again in at most C periods and stay there. Moreover, for this to happen, *some* member i of S must participate in some profitable move z from y (for if not, all members of S must receive $\mathbf{u}'$ for ever after, a contradiction to the fact that x is the unique limit).

This means that

(13.2) $V_i(z,p) \geq V_i(y,p).$

Now observe that

$$V_i(y,p) = (1 - \delta_i)u_i' + \delta_i V_i(z,p),$$

so that

$$\begin{aligned}
V_i(z,p) - V_i(y,p) &= (1 - \delta_i)[V_i(z,p) - u_i'] \\
&\leq (1 - \delta_i)\left[(1 - \delta_i^C)M + \delta_i^C u_i - u_i'\right] \\
&< \epsilon + u_i - u_i' \\
&\leq 0.
\end{aligned}$$

But this inequality contradicts (13.2).   □

Proof of Proposition 13.4. Let C be the total number of states. Let M and W be the maximal and minimal (one-period) payoffs to any player. Pick $\delta^* \in (0,1)$ such that for any two states x and y in X and any index i with $u_i(x) > u_i(y)$, we have (i) $u_i(x) > (1 - \delta^{*C})M + \delta^{*C}u_i(y)$, and (ii) $(1 - \delta^{*C})W + \delta^{*C}u_i(x) > u_i(y)$. Consider any collection of discount factors all in $(\delta^*, 1)$, and fix some absorbing deterministic EPCF. Let $Z \subseteq X$ be its set of absorbing states.

Let $z \in Z$ be some absorbing state. Fix any coalition S and consider any $x \in F_S(z)$. Use the notation $x^0, x^1, \ldots, x^m$ to describe the subsequent path prescribed by the PCF starting from $x = x^0$ and ending at the absorbing state $x^m = y \in Z$. Because the PCF is an equilibrium, we also know that there are coalitions $S_0, S_1, \ldots, S_{m-1}$ such that for $j = 0, \ldots, m - 1$, $x^{j+1} \in F_{S_j}(x^j)$ and

(13.3) $V_i(x^{j+1}) \geq V_i(x^j)$

for all $i \in S_j$. Now observe that $V_i(x^j) = (1 - \delta_i)u_i(x^j) + \delta_i V_i(x^{j+1})$, so that by (13.3),

(13.4) $V_i(x^{j+1}) \geq u_i(x^j)$

for each S_j and $i \in S_j$. Next, note that

$$V_i(x^{j+1}) \leq (1 - \delta_i^C)M + \delta_i^C u_i(y)$$

(because the PCF from x^{j+1} leads to the absorbing state y in at most C steps), and combining this with (13.4), we may conclude that

$$(1 - \delta_i^C)M + \delta_i^C u_i(y) \geq u_i(x^j).$$

But this means (by (i) in our definition of δ^*) that

(13.5) $$u_i(y) \geq u_i(x^j)$$

for all S_j and all $i \in S_j$. (13.5) proves that y, apart from being in Z, sequentially dominates x.

Moreover, since x is a possible move (by S) from z and z is an absorbing state, $V_i(x) \leq V_i(z)$ for some $i \in S$. Because z is absorbing, we know that $V_i(z) = u_i(z)$, so that

(13.6) $$V_i(x) \leq u_i(z).$$

Now observe that

$$V_i(x) \geq (1 - \delta_i^C)W + \delta_i^C u_i(y)$$

(because the PCF from x leads to the absorbing state y in at most C steps). Combining this with (13.6) we may conclude that

$$u_i(z) \geq (1 - \delta_i^C)W + \delta_i^C u_i(y)$$

By part (ii) in the definition of δ^*, we deduce that

(13.7) $$u_i(z) \geq u_i(y)$$

for some $i \in S$.

Now (13.5) and (13.7) together prove that $f(Z) \supseteq Z$, where $f(Z)$ is the set of all states x such that for every coalition S and for any state $y \in F_S(x)$, there exists $z \in Z$ (where either $y = z$ or z sequentially dominates y) such that the inequality $u_i(x) \geq u_i(z)$ holds for some $i \in S$. Using the same argument in the proof of Proposition 1 in Chwe (1994), we may conclude that Z is contained in the largest consistent set. □

Proof of Proposition 13.5. First we prove that $p(\mathbf{a}^*, \mathbf{a}^*) = 1$. Suppose not. Then there is some coalition S and a move to a state $\mathbf{a}$ such that $V_i(\mathbf{a}, p) \geq V_i(\mathbf{a}^*, p)$ for all $i \in S$. Let $\mathbf{a}'$ be some state having the lowest value of V_i among all states satisfying the requirement of the previous sentence. (By the assumption of common payoffs and a common discount factor, the same state can achieve this for every player.) By not moving, each member i of coalition S gets a payoff of

$$(1 - \delta)u_i(\mathbf{a}^*) + \delta \sum_{\mathbf{a} \in A} p(\mathbf{a}^*, \mathbf{a})V_i(\mathbf{a}, p)$$

which is obviously larger than $V_i(\mathbf{a}', p)$, a contradiction.

Next, we show that $\mathbf{a}^*$ is the unique absorbing limit. To this end, we first note that if $\mathbf{a}' \neq \mathbf{a}^*$, then there exists $\mathbf{a}'' \neq \mathbf{a}'$ such that

$p(\mathbf{a}', \mathbf{a}'') > 0$. Suppose not; then $p(\mathbf{a}', \mathbf{a}') = 1$ for some $\mathbf{a}' \neq \mathbf{a}^*$. In particular, $V_i(\mathbf{a}', p) = u_i(a')$, while $V_i(\mathbf{a}^*, p) = u_i(\mathbf{a}^*)$. However, since $u_i(\mathbf{a}^*) > u_i(\mathbf{a}')$, we have $V_i(\mathbf{a}^*, p) > V_i(\mathbf{a}', p)$ for any $i \in N$. So there is a strictly profitable move from $\mathbf{a}'$, which contradicts requirement (ii) of an EPCF. This establishes the claim at the beginning of the paragraph.

Now, if $\mathbf{a}^*$ is not the unique absorbing limit, then the set $C(p) \equiv \{\mathbf{a} \in A : \text{for any } k \geq 1, p^{(k)}(\mathbf{a}, \mathbf{a}^*) = 0\}$ is nonempty. By the common payoff assumption, there exists $\mathbf{a}' \in C(p)$ such that $V_i(\mathbf{a}', p) \geq V_i(\mathbf{a}, p)$ for any $\mathbf{a} \in C(p)$. By the claim of the previous paragraph, there is a state $\mathbf{a}''$ such that $p(\mathbf{a}', \mathbf{a}'') > 0$.

In order to satisfy requirement (i) of an EPCF, it must be (recalling common payoffs) that $V_i(\mathbf{a}', p) \leq V_i(\mathbf{a}'', p)$ for all i. But it is obvious that $\mathbf{a}'' \in C(p)$. Consequently, from the definition of $\mathbf{a}'$ it follows that $V_i(\mathbf{a}', p) = V_i(\mathbf{a}'', p)$, and indeed, this is true for any state $\mathbf{a}''$ such that $p(\mathbf{a}', \mathbf{a}'') > 0$.

At the same time, we know that $V_i(\mathbf{a}^*, p) = u_i(\mathbf{a}^*) > V_i(\mathbf{a}', p)$. Therefore, we conclude that although $V_i(\mathbf{a}^*, p) > V_i(\mathbf{a}', p)$ for any $i \in N$ (common payoffs), $p(\mathbf{a}', \mathbf{a}'') > 0$ occurs only for $\mathbf{a}''$ with $V_i(\mathbf{a}'', p) = V_i(\mathbf{a}', p)$. This violates requirement (ii) of an EPCF, a contradiction.   □

13.7 Summary

This chapter studies a "real-time" model of coalition formation. As in Part 2 of this book, such a process is viewed as a (possibly) stochastic sequence in which coalitions and payoffs evolve over time. However, in contrast to the protocol-based bargaining model studied in Part 2, the approach in this chapter is in the tradition of cooperative game theory. The process of coalition formation is spurred on by coalitional moves to a new state. Such moves are not explicitly proposed by any individual; they are simply presumed to occur only if all individual members of the coalition gain thereby.

I "benchmark" the equilibrium concept by using a special case of it to characterize the core. More precisely, I study the class of deterministic processes of coalition formation with unique absorbing limit. I show that in all models of coalition formation that are derived from an underlying characteristic function, the limits of such processes (essentially) coincides with the core (Propositions

13.2 and 13.3), provided that discount factors are close enough to unity. Apart from benchmarking our solution concept, this result is of independent interest because it reveals an interesting consistency property of the core, which goes beyond the well-known "internal consistency" of the core; see Observation 11.1.[20]

Next, I consider deterministic schemes that do not necessarily have a unique limit (but nevertheless do not display cycles). I show by means of an example that non-core limits might now emerge. However, it turns out that such schemes yields absorbing states that always lie within the "largest consistent set" (Chwe (1994)), provided that discount factors are close enough to unity.

This result (Proposition 13.4) is valid without any restrictions on the underlying model of coalition formation and serves as a second important benchmark. However, the inclusion of absorbing limits within the largest consistent set is generally strict. Example 13.3, which shows this, brings out the fact that our solution concept imposes more restrictions on the final outcomes than the largest consistent set does. But this does not mean that the outcomes selected by our solution are necessarily the "more efficient" ones; Example 13.4 is devoted to an understanding of this point.

Next, I make some observations on deterministic cyclical solutions. These typically exist in situations in which core-like restrictions lead to an empty outcome. But there are examples in which no such solution (and indeed, no deterministic solution) exists. This motivates a study of probabilistic solutions, which takes up the final section of the chapter.

Uncertainty enters a process of coalition formation in two possible ways. First, a particular coalition may be able to induce two or more states which are not payoff-comparable, and might randomize (or be perceived as randomizing). Second, it is possible that at some state several coalitions have access to profitable moves, and that these are chosen randomly.

---

[20]Green (1974) and Sengupta and Sengupta (1996) prove a related result: starting from an arbitrary state, a sequence of profitable coalitional deviations lead to a core state (Green shows this for exchange economies and Sengupta and Sengupta for TU games). But there is a fundamental difference between these results and the one established here. The earlier results assume that players are myopic, so that members of a moving coalition do not foresee what happens after their immediate deviation takes place. In the model described here, in contrast, individuals are farsighted and will need to forecast future deviations or moves.

It turns out that such forms of randomization occur naturally in strategic form games, in the sense that randomization is often necessary for existence of an equilibrium process (contrast this with characteristic functions). Accordingly, we focus in the section on games in strategic form. The simplest (though by no means trivial) starting point is games with common payoffs. We show that for such games, every equilibrium must lead to the efficient outcome, provided that discount factors are close to unity (Proposition 13.5).

But this result fails when we depart from common payoffs. For instance, we show (Example 13.6) that a 2×2 symmetric coordination game may generate equilibria that hone in on the "bad" equilibrium. The stochastic nature of the equilibrium is explained in detail; indeed, we argue that such an equilibrium *must* be stochastic.

We turn finally to a detailed analysis of the Prisoners' Dilemma. Our solution concept applied here yield a rich variety of outcomes (though, to be sure, not everything is possible). The main points are: (1) cooperation can be sustained using deterministic schemes, while defection can never be sustained in this way (provided that discount factors are close to unity); (2) in contrast, stochastic schemes can support defection as an absorbing state, and can also generate cycles of movement with possibly some inertia at the cooperative outcomes; and (3) cardinalities do matter in pinning down equilibria — Observations 13.2 and 13.3 and Examples 13.7–13.9, which conclude the chapter, make these points amply clear.

Appendix A

Remarks on Efficient Moves.

We make some brief remarks on the notion of efficient moves in the definition of an EPCF. As observed in the main text, one might weaken the definition of an EPCF to allow for all profitable moves, not just the efficient ones. Call such an EPCF a weak EPCF.

1. The existence of weak EPCFs is obviously not an issue, because an EPCF is clearly a weak EPCF.

2. Weak EPCFs might lead to outcomes that appear unreasonable. Consider the following example with two players — 1 and 2 — and three states, x, y and z. Payoffs are as follows:

	x	y	z
1	0	1	100
2	1	0	100

Suppose further that individual 1 can induce y and z from x, while 2 can induce x and z from y, and that no other coalition/state combination permits nontrivial moves.

Now it is easy to construct a weak EPCF (for all discount factors, in fact) in which, starting from either x or y, the system endlessly oscillates between x and y, *even though* either player could induce z and make both players much better off. An EPCF would negate this possibility by permitting — indeed, *demanding* — that each player make an efficient move.

3. At the same time, it is worth noting that our core characterization theorems may be strengthened by taking note of the distinction between EPCFs and weak EPCFs. This is true in the following sense. In Proposition 13.2, a (strong) core outcome is "implemented" by an EPCF (satisfying the efficient moves principle). At the same time, a cursory glance at the proof of Proposition 13.3 will reveal that every weak deterministic EPCF with unique absorbing limit must pick out a (weak) core allocation. That is, Proposition 13.3 applies to the broader class of weak EPCFs.

4. It hardly needs to be mentioned that the "efficient moves" requirement need not lead to efficiency overall, for exactly the same reason that Nash equilibria need not be Pareto optimal. For instance, Observation 13.3 tells us that an EPCF may lead to mutual defection as its unique absorbing limit in the case of the Prisoners' Dilemma.

5. However, there is an important sense in which our equilibrium concept fails to capture certain aspects of "efficient moves". We have proceeded entirely in the spirit of dynamic games, in which the one-shot deviation principle is applied: players take not only the strategies of other players as given, they take as given their *own* strategies in the future. By the well known principle of Blackwell that "unimprovability implies optimality" in discounted situations, there is seemingly no loss of generality in doing this.

But the principle fails when "players" are coalitions (and also if players have vector-valued objectives). To create a profitable deviation for the coalition as a whole (or for every component of the vector payoff function), several moves may be needed. Indeed, this sort of consideration lies behind the efficiency failure in Example 13.4. There, the coalition {12} *can* engineer, if it so wishes, a move from one of the inefficient absorbing states a or c. However, a move from a only ends at c, and *vice versa*, so that both players cannot find it simultaneously worthwhile to participate in the proposed move. At the same time, if players {12} were to simultaneously deviate at *both* a and c, the "double deviation" would indeed be worthwhile.

This raises a conceptual issue. The principle of one-step deviations is built into our solution concept: individuals and coalitions at different dates are regarded as different individuals and coalitions. Therefore coalitions are as involved (in this conceptualization) in a game against themselves as against other coalitions. It is unclear whether this formulation should be dropped (compare this, for instance, with the literature on changing preferences, e.g., Strotz (1958) and Phelps and Pollak (1968)). We tentatively retain it, despite the disturbing feature of Example 13.4.

6. Finally — while accepting the efficient moves principle — one might question the particular formulation adopted in our definition. For instance, one could rule out an efficient move for S (as described by us) if there is some strict subset of S, say T, which can generate *another* change that makes *its* members still better off relative to the payoff under the efficient move (by S). In this case one might want to assign probability zero to the move by S (and positive probability to the move by T). However, this refinement raises other issues. One interpretation of the probabilistic nature of a move is that Nature chooses a coalition randomly and permits it to enjoy a profitable deviation. In that case, the subset T might be bound by the decisions of the entire coalition S. On the other hand, if this interpretation is rejected, then other problems arise. For instance, why restrict the search for better moves to subsets of S and not other sets T which share a common intersection with S (where the intersecting members are allowed to go with the coalition that has the better move)? But this further refinement leads to possible circularities, rendering a conceptually satisfactory definition impossible. At the same time, it should be noted that such potential circularities in defining efficient moves — which we avoid by assumption — do not in any way preclude the study of cycles over time, which are allowed for in the definition.

Appendix B

Proofs Omitted in the Main Text.

Proof of Proposition 13.1. Denote by $\mathcal{P}$ the set of all possible PCF's. We construct a correspondence $\phi : \mathcal{P} \Rightarrow \mathcal{P}$, show that a fixed point exists, and observe that a fixed point of ϕ must be an EPCF.

We begin by observing that for every $p \in \mathcal{P}$, a unique value function $V_i(x,p)$ exists for each player i, satisfying (13.1). Let $\mathbf{V}_i(p)$ denote the vector of payoffs $\{V_i(x,p)\}_{x \in X}$, $\mathbf{u}_i$ the vector of current payoffs $\{u_i(x)\}_{x \in X}$, and P the matrix of transition probabilities (under p). Then (13.1) may be immediately rewritten as

$$(I - \delta_i P)\, \mathbf{V}_i(p) = \mathbf{u}_i.$$

Since $\delta_i \in (0,1)$, $I - \delta_i P$ has a dominant diagonal. This guarantees the unique solvability and continuity of $\mathbf{V}_i(p)$ in p.

To construct ϕ, first consider (x, p) such that strictly profitable moves exist; let $Y(x, p)$ be the set of all strictly profitable and efficient moves. For each $y \in Y(x, p)$ there is a coalition S such that y is strictly profitable and efficient for S from x (under p). Call such a coalition *allowable* (given (y, x, p)), and for each allowable coalition S define $\sigma_S(y, x, p) \equiv \min_{i \in S}[V_i(y, p) - V_i(x, p)]$. Having done so, let $\sigma(y, x, p) \equiv \max_S \sigma_S(y, x, p)$, where the maximum is taken over allowable coalitions S. Now define a probability measure over $Y(x, p)$ — call it $q(x, p)$ — by

(13.8)
$$q(x, p)[y] \equiv \frac{\sigma(y, x, p)}{\sum_{y' \in Y(x,p)} \sigma(y', x, p)}.$$

Define a correspondence $\Delta(x, p)$ as follows: when strictly profitable moves exist, $\Delta(x, p) = \{q(x, p)\}$. Otherwise, $\Delta(x, p)$ be the collection of *all* probability measures with support contained in the union of $\{x\}$ and the collection of weakly profitable and efficient moves from x (under p).

Obviously, $\Delta(x, p)$ is nonempty and convex-valued for each (x, p). Now we claim that it is uhc in p for given x. To this end, let p^k be some sequence in $\mathcal{P}$ converging to p. Study a corresponding sequence $q^k \in \Delta(x, p^k)$ and extract a convergent subsequence converging to some q (retain original sequence notation). We claim that $q \in \Delta(x, p)$.

This claim is obviously true if no strictly profitable move exists at (x, p).[21] So suppose that a strictly profitable move does exist at (x, p). We note that for any $y \in Y(x, p)$, $\sigma(y, x, p^k) \to \sigma(y, x, p)$ as $k \to \infty$. (This is very easy to verify, using the fact that $V_i(x, p)$ is continuous in p for every i and x.)

In particular, this means that for k large enough, $\Delta(x, p^k)$ is a singleton containing the probability measure $q(x, p^k)$ defined by (13.8). It also means that $q(x, p^k) \to q(x, p)$.

We have therefore shown that $\Delta(x, p)$ is nonempty, convex-valued and uhc in p for each x. Define $\phi : \mathcal{P} \Rightarrow \mathcal{P}$ by $\phi(p) = \prod_{x \in X} \Delta(x, p)$ for every $p \in \pi$. Then, by the arguments above, all the conditions for the Kakutani fixed point theorem are satisfied, and there exists $p^* \in \mathcal{P}$ such that $p^* \in \phi(p^*)$. It is easy to see that p^* satisfies all the conditions of an EPCF. □

---

[21]All we need to observe is that if y is strictly profitable for the sequence (x, p^k), then it must be weakly profitable for (x, p).

Proof of Observation 13.1. I first construct an equilibrium in which every state x' either moves to state y' with probability p, where $(x', y') \in \{(x, y), (y, z), (z, x)\}$, or continues unchanged with probability $1 - p$. Then

$$
\begin{pmatrix}
1 - \delta + \delta p & 0 & -\delta p \\
-\delta p & 1 - \delta + \delta p & 0 \\
0 & -\delta p & 1 - \delta + \delta p
\end{pmatrix}
\begin{pmatrix}
V^H \\
V^M \\
V^L
\end{pmatrix}
= (1 - \delta)
\begin{pmatrix}
1 \\
a \\
0
\end{pmatrix}
$$

where

$$
\begin{pmatrix}
V^H \\
V^M \\
V^L
\end{pmatrix}
=
\begin{pmatrix}
V_1(x, p) \\
V_1(z, p) \\
V_1(y, p)
\end{pmatrix}
=
\begin{pmatrix}
V_2(y, p) \\
V_2(x, p) \\
V_2(z, p)
\end{pmatrix}
=
\begin{pmatrix}
V_3(z, p) \\
V_3(y, p) \\
V_3(x, p)
\end{pmatrix}.
$$

By solving this equation we see that

$$
V^H = D\{(1 - \delta + \delta p)^2 + a(\delta p)^2\}
$$
$$
V^M = D\{a(1 - \delta + \delta p)^2 + \delta p(1 - \delta + \delta p)\}
$$
$$
V^L = D\{(\delta p)^2 + a\delta p(1 - \delta + \delta p)\},
$$

where $D = (1 - \delta)/[(1 - \delta + \delta p)^3 - (\delta p)^3] > 0$, and p denotes the probability of moving to the next state.

Note first that $V^M - V^L = D(1 - \delta)[\delta p + a(1 - \delta + \delta p)] > 0$. Thus, a player who is currently getting 0 surely joins a coalitional move. The question is whether a player who is currently getting a would also do so. Some tedious calculations show that

$$
V^H - V^M = D(1 - \delta)\{(1 - \delta + \delta p) - a(1 - \delta + 2\delta p)\},
$$

so that

$$
V^H \geq (<)V^M \iff a \leq (>)\frac{1 - \delta + \delta p}{1 - \delta + 2\delta p}.
$$

Note that $p = 1$ if $V^H > V^M$, and $p \in (0, 1)$ can occur only if $V^H = V^M$. Thus, when $a < 1/(1 + \delta)$, $V^H > V^M$ holds for any p, and we must conclude that $p = 1$. Similarly, when $a = 1/(1 + \delta)$, the only possibility is, again, $p = 1$.

When $a > 1/(1+\delta)$, p can no longer be 1, since $p = 1$ implies $V^H < V^M$. Since $a < 1$, neither can it be that $p = 0$ ($V^H > V^M$). Thus, the only possibility left is the case where $V^H = V^M$ so that $p \in (0, 1)$ holds. Hence, when $a > 1/(1 + \delta)$, $p = (1 - a)(1 - \delta)/\delta(2a - 1)$ is the unique symmetric EPCF.

The rest of the proof shows that no deterministic EPCF exists when $a > 1/(1 + \delta)$. Suppose one does. First consider the case in which $p(x', x') = 0$ for all states x'. In this case, $p(x, y) = 1$. For if not, we have $p(x, y) = 0$, so that $p(x, z) = 1$. But then it is easy to see that player 1 will never agree to the move from x to z. By the same argument applied to each state, we conclude that $p(x, y) = p(y, z) = p(z, x) = 1$. But we've argued above that this is impossible when $a > 1/(1 + \delta)$.

So the only remaining case is one in which $p(x', x') = 1$ for some state, say $x' = x$. In this case we must have $p(y, y) < 1$ (otherwise {23} will have a strictly profitable move from x) and $p(z,z) < 1$ (because {12} has a strictly profitable move from z). By determinism, we may conclude that $p(y,y) = p(z,z) = 0$. Now, $p(y,x)$ must be 0 as well (2 would not agree to move from y under a deterministic EPCF), so $p(y,z) = 1$. Likewise, we have $p(z,y) = 0$ (3 will not agree to move under a deterministic EPCF), so $p(z,x) = 1$. It follows that for player 3,

$$V_3(z,p) = 1 - \delta,$$

while by refusing to move from y under the deterministic EPCF, player 3 can guarantee herself a value of a. Consequently, $V_3(y,p) \geq a$. It is easy to see that under our assumptions, $a > 1 - \delta$. We may therefore conclude that $V_3(y,p) > V_3(z,p)$, which contradicts the fact that coalition {23} moves the state from y to z. □

Proof of Observation 13.2. First, note that in any deterministic EPCF, we must have $p(y,y) = p(z,z) = 0$ and $p(y,w) = p(z,w) = 1$.

Case a. Deterministic EPCFs with unique absorbing state x. We already know that $p(y,w) = p(z,w) = 1$ as described above. Moreover, it must be that $p(w,x) = 1$. Notice that $V_i(w,p) = \delta V_i(x,p) > 0$, so $V_i(w,p) < V_i(x,p)$, so the move from w to x is strictly profitable for both players. It remains to check that a unilateral move from x to, say, y is not worthwhile. Such a move fetches player 1 $(1-\delta)a + \delta V_i(w,p) = (1-\delta)a + \delta^2 V_i(x,p) = (1-\delta)a + \delta^2$, whereas remaining at x yields 1. It follows that such a deterministic EPCF exists if and only if $a \leq 1/(1+\delta)$.

Case b. Deterministic EPCFs with unique absorbing state w. Once again, we know that $p(y,w) = p(z,w) = 1$, and also that $p(w,w) = 1$. Because $p(x,x) = 0$, consider first the possibility that $p(x,w) = 1$. This isn't possible because — using determinism — either player can negate such a move and guarantee a flow of 1 forever. Therefore, I analyze a PCF with $p(x,y) = 1$. (Since it is deterministic, the EPCF must treat moves to y and z asymmetrically. The only other case is the mirror image of this, to be studied in exactly the same way.)

First I justify $p(x,y) = 1$. For player 1 to move from x, we need $(1-\delta)a + \delta V_1(w,p) \geq V_1(x,p)$. Note that $V_1(x,p) = (1-\delta) + \delta(1-\delta)a + \delta^2 V_1(w,p)$ and $V_1(w,p) = 0$. Thus, what we need is $a \geq 1 + \delta a$ or $a \geq \frac{1}{1-\delta}$.

Second, we must "justify" $p(w,w) = 1$. It is necessary and sufficient to show that player 2 (who suffers more by moving to x) does not want to do so. This condition is $0 = V_2(w,p) \geq V_2(x,p) = 1 + \delta b + \delta^2 V_2(w,p) = 1 + \delta b$, or $\delta b \leq -1$. These two conditions on a and b are precisely those listed in the statement of the observation.

Case c. Deterministic cyclical EPCFs. Once again, there are two such solutions, to be treated entirely symmetrically: one with $p(x,y) = 1$ and the other with $p(x, z) = 1$. We study the case in which $p(x, y) = p(y, w) = p(z, w) = p(w, x) = 1$. To assure ourselves that $p(x, y) = 1$, it is necessary and sufficient that $a + \delta V_1(w, p) = a + \delta^2 V_1(x, p) \geq V_1(x, p)$, or $a \geq (1 - \delta^2) V_1(x, p)$. But it is easy to see that

$$V_1(x, p) = (1 + \delta a + \delta^2 0) + \delta^3 (1 + \delta a + \delta^2 0) + \ldots = \frac{1 + \delta a}{1 - \delta^3},$$

so that our required condition is

$$a \geq (1 - \delta^2) \frac{1 + \delta a}{1 - \delta^3},$$

which is equivalent to $a \geq 1 + \delta$.

To make sure that $p(w, x) = 1$, it is necessary and sufficient to check incentives for the "weaker" player 2: $V_2(x, p) \geq V_2(w, p)$, or equivalently, $V_2(x, p) \geq 0$. Direct computation tells us that

$$V_2(x, p) = (1 + \delta b + \delta^2 0) + \delta^3 (1 + \delta b + \delta^2 0) + \ldots = \frac{1 + \delta b}{1 - \delta^3},$$

so that the condition in question is simply $b \geq -\frac{1}{\delta}$. This concludes the study of all deterministic EPCFs. ☐

Proof of Observation 13.3.

Case a. Stochastic symmetric EPCFs with an absorbing state w. To display these, we set up the relevant value functions. By symmetry, we can simply focus on Player 1:

$$V_1(x, p) = 1 + \frac{\delta}{2} (V_1(y, p) + V_1(z, p)),$$
$$V_1(y, p) = a + \delta V_1(w, p),$$
$$V_1(z, p) = b + \delta V_1(w, p),$$
$$V_1(w, p) = 0.$$

By substituting $V_1(w, p) = 0$ in the other equations, we may conclude that $V_1(y, p) = a$, $V_1(z, p) = b$, and

$$V_1(x, p) = 1 + \frac{\delta}{2} (a + b).$$

Since $a < 0$ and $b < 0$, $p(y, w) = p(z, w) = 1$ are incentive compatible. Moreover, $p(w, w) = 1$ can be supported if $V_1(x, p) \leq 0$, or $a + b \leq -\frac{2}{\delta}$. The question is whether we can support $p(x, y) = \frac{1}{2}$. The answer is yes if $V_1(y, p) \geq V_1(x, p)$, or $a \geq \frac{2}{2 - \delta} + \frac{\delta}{2 - \delta} b$. Thus, if (i) $a + b \leq -\frac{2}{\delta}$, and (ii) $a \geq \frac{2}{2 - \delta} + \frac{\delta}{2 - \delta} b$, then the above PCF is indeed an EPCF. Since condition (ii) is satisfied trivially (a is the highest payoff of all), condition (i) is the only one that needs to be satisfied. That is, if $a + b \leq -\frac{2}{\delta}$, then w can be supported as the unique absorbing state of a symmetric stochastic EPCF.

Case b. Stochastic symmetric cyclical EPCFs. We analyze a PCF with $p(x, y) = p(x, z) = p \le \frac{1}{2}$, and $p(y, w) = p(x, w) = p(w, x) = 1$. Again, the key conditions are (i) $V_1(y, p) \ge V_1(x, p)$, and (ii) $V_1(x, p) \ge V_1(w, p)$. It is easy to see that

$$
\begin{aligned}
V_1(x, p) &= 1 + \delta\{p(V_1(y, p) + V_1(z, p)) + (1 - 2p)V_1(x, p)\} \\
&= 1 + \delta p(a + b) + 2\delta^2 p V_1(w, p) + \delta(1 - 2p)V_1(x, p) \\
&= 1 + \delta p(a + b) + 2\delta^3 p V_1(x, p) + \delta(1 - 2p)V_1(x, p),
\end{aligned}
$$

so that

$$
\{1 - \delta + 2\delta p - 2\delta^3 p\}V_1(x, p) = 1 + \delta p(a + b),
$$

or equivalently,

$$
V_1(x, p) = \frac{1 + \delta p(a + b)}{(1 - \delta)(1 + 2\delta p(1 + \delta))}.
$$

Now, we are ready to check condition (i). We need $V_1(y, p) = a + \delta^2 V_1(x, p) \ge V_1(x, p)$, or $\frac{a}{1 - \delta^2} \ge V_1(x, p)$. Thus, we need

$$
\frac{1 + \delta p(a + b)}{(1 - \delta)(1 + 2\delta p(1 + \delta))} \le \frac{a}{(1 - \delta)(1 + \delta)}.
$$

Hence, condition (i) boils down to

$$
b \le \{\frac{1}{(1 + \delta)\delta p} + 1\}a - \frac{1}{\delta p}.
$$

Now for condition (ii). This is equivalent to $V_1(x, p) \ge 0$, or $a + b \ge -\frac{1}{\delta p}$. Putting these two conditions together, we finally obtain:

(13.9) $$ -a - \frac{1}{\delta p} \le b \le \{\frac{1}{(1 + \delta)\delta p} + 1\}a - \frac{1}{\delta p}. $$

Recall $p \in (0, \frac{1}{2}]$. Thus, an stochastic EPCF with $p(x, x) = 0$ ($p = \frac{1}{2}$) can be supported in the parameter range of $-a - \frac{2}{\delta} \le b \le (\frac{2 + \delta + \delta^2}{\delta + \delta^2})a - \frac{2}{\delta}$. The second inequality is almost always satisfied if $a > 1$ (the Prisoners' Dilemma) and if δ close to 1. The first inequality is more demanding and we need $-2 \le a + b$ when δ is close to unity.

What if p is less than $\frac{1}{2}$? In this case, we have $p(x, x) > 0$ holds, which means that players 1 and 2 need to be indifferent between deviating from x and staying at x. This means that the second inequality in (13.9) must hold with equality. This implies that

$$
p = \frac{1 - \frac{a}{1 + \delta}}{\delta(a - b)}.
$$

Thus, if $a \ge 2$, then there is no such $p > 0$ for any $\delta < 1$, and we can only have a stochastic EPCF with $p = \frac{1}{2}$. If $a < 2$, then the above p satisfies the first inequality as long as $\delta > a - 1$, and $p < \frac{1}{2}$ can form a stochastic EPCF. □

CHAPTER 14

Directions

In this book, I've tried to outline an approach to coalition formation and binding agreements, one that adopts an explicitly game-theoretic perspective. It goes without saying that a game-theoretic approach to this subject isn't novel at all. Some of the earliest questions in game theory were about binding agreements, as the magisterial tome by von Neumann and Morgenstern (1944) so clearly reveals. In this sense, I simply continue in the tradition of a long literature.

Two methodological themes receive emphasis throughout this book. First, while an agreement, *once agreed to*, may be implemented at little or no cost, the process by which that agreement is arrived at is fundamentally noncooperative. By marrying noncooperative game theory to the traditional structure of cooperative games, we can gain insight into this process. Second, I highlight throughout a particularly significant failure of cooperative game theory. In its near-universal acceptance of the characteristic function, it has unwittingly downplayed the immense importance of externalities across formed coalitions. (The characteristic function essentially assumes those externalities away by assigning an unambiguous worth to each coalition, irrespective of the ambient coalition *structure* in which that coalition is located.)

These two themes combine in interesting ways. For more on that you will have to read the book, if you haven't done so already. However, one overriding implication, which requires further research, is that coalitional negotiations are often grossly inefficient. We've been brought up on a steady diet of incomplete information — or failure of contractibility — to understand the inefficiencies that we see around us. And of course, that viewpoint has yielded deep and useful insights. Yet — and this is even more true of someone brought up in a developing country — there is a huge variety of situations in which incompleteness in information may not represent the appropriate first cut. In the debates around major issues — protectionism, infrastructural investments such as large dams, land reform, taxation, the transition from agriculture to industry, ethnic conflict — the vested interests of different parties are not exactly state secrets. The Coase theorem would have us contract these problems away, moving inexorably at each instance to the the surplus maximizing outcome. That doesn't happen.

To be sure, common sense dictates that no one approach can do justice to these issues. Incomplete information, the lack of commitment, and the difficulty of internalizing diffuse costs and benefits all have their role to play. My only objective here is to also call attention to the process of coalition formation, stripped of all these other features that we understand relatively well. The hope is that such theories will also become first-order participants in our understanding of the obstacles to sustained economic development.

This monograph represents a particular perspective on coalition formation, and in no way attempts to be a comprehensive treatise on the subject. At the same time, lack of comprehensiveness cannot be an excuse for ignoring complementary avenues of progress and new directions. In this final chapter I'd like to explore some of these themes. The classification of the material in this chapter is entirely whimsical. Some of the headings are methodological; others concern new questions or topics. I hope that the reader will forgive this mixed-up taxonomy.

14.1 Coalition Formation Without Unanimity

An assumption maintained throughout this monograph is that *every* coalition member must agree in the formation of a coalition. In many situations — political legislation comes to mind as an example — this is a strong restriction. Majority approval or more generally, the achievement of some given supermajority, may be enough to implement an outcome. Baron and Ferejohn (1989) formally develop a noncooperative bargaining game along the lines of the Rubinstein–Ståhl model with unanimity replaced by majority.

We've already discussed the "closed rule" version of the Baron–Ferejohn model in Section 4.5 of Chapter 4. A cake of unit size is to be divided among n players, who make proposals. Responses are sequential, but only some given supermajority $m > n/2$ need approve the proposal for it to be implemented. Once a proposal is rejected a fresh proposer is chosen at random. Once a proposal is accepted the game is over.

We've seen in Section 4.5 that such a model of bargaining with supermajority approval is equivalent to the study of *unanimous* bargaining for the characteristic function

$$v(S) = 1 \text{ if and only if } |S| \geq m.$$

It therefore predicts the formation of minimal winning coalitions of size m. Of course, a suitably defined characteristic function bargaining model with unanimity would yield exactly the same result.

For games with ongoing negotiation, such as those introduced in Chapter 8, it is unclear how to study nonunanimity protocols. Recall our study of binding agreements there: we presumed that ongoing agreements are in force unless unanimously relinquished by those whose coalitional memberships are altered (by the fresh proposal under consideration). If we adopt the interpretational view, described above, that the agreement of a subcoalition S is simply a proxy for a majority agreement within a larger coalition (which is the grand coalition in the case of the Baron–Ferejohn model, but need not be so more generally), then who is it that must approve the dissolution of that agreement? Must the breaking up of an existing agreement be settled by unanimity? Or if not, must a majority of the affected players agree, or simply a majority of players directly involved in the fresh initiative? Questions such as these must often be addressed on a case-by-case basis. It is hoped that the general methodology developed for reversible agreements remains largely valid, but this will require more careful study.

14.2 Equity Within Coalitions

The Baron–Ferejohn model contains a second implication, one that is fundamentally the product of the random-proposer protocol. It is that winning coalitions will generally not share the surplus equally, and that the proposer (within the winning coalition) will pick up a larger share. When the protocol is "rejector-proposes", the winning coalition must exhibit approximately equal division, as per our results in Chapter 5, even though only a majority is required for implementation.

This sort of unequal division stems from two sources; first, that the ability to make a proposal allows the proposer to be part of some advantageous subcoalition (a winning coalition in the special case of the Baron–Ferejohn model); and second, that the mantel of proposer is not readily granted to a responder who rejects, a feature of the random-proposer model.[1] These two forces create a situation in which the responder is fundamentally weaker than the proposer, not because of any innate ability or type differences, but simply because of bargaining protocol.

In this book, we've chosen instead not to focus too strongly on these differences. If someone is included in a proposal, then — at least from the bargaining perspective alone — she is roughly on equal terms with the proposer. This is why our bargaining model tends to generate equal division within coalitions. To be sure, someone may be excluded from the coalition altogether, and then there is no presumption of equal division between the "insider" and the "outsider".

Certainly, our model allows for unequal division within coalitions when the agents in those coalitions have truly different characteristics or outside options. This is what happens in the more general coalitional bargaining model studied in Chapter 7, but nevertheless there is a tendency towards "as much equality as possible", subject to outside options (see the discussion in Section 7.4.4).

My focus on such equilibria stems from an underlying view — one that may be worth more systematic exploration — that within-group equity among individuals viewed as *ex ante* similar is a very strong driving force of human interaction. Within any group, acknowledged to be engaged in some joint enterprise, the forces making for equal treatment are strong (unless some special member is universally acknowledged to have superior outside options or abilities). In contrast, wholesale exclusion — the forming of separate

---

[1]As we've seen in our discussion of the Rubinstein–Ståhl bargaining model (see Section 4.4), both these conditions are really required. In that model, there is only one coalition to belong to — the grand coalition — and the random-proposer model cannot prevent equal division as discounting vanishes.

coalitions — rarely appears to be a problem, and questions of equity arise less often.

In-group equity concerns appear to be fairly ubiquitous in experimental studies of bargaining. It is well known that even in simple bargaining situations in which a proposer has *all* the power, such as in the ultimatum game, a high degree of equity is preserved in observed divisions of the surplus. Several authors have attempted to explain such outcomes by invoking notions of fair play on the part of the proposer or anger/spite on the part of the responder if the proposer abuses his privileged position.

There is far less on in-groups versus out-groups in a coalitional bargaining context. But a recent paper by Frechette, Kagel and Morelli (2005) [FGM] that is particularly sensitive to this question finds strong evidence to support the in-group/out-group hypothesis. Among other things, the paper serves as an experimental assessment of the Baron–Ferejohn model. The authors consider a 5-person model of majority bargaining in which the players need to divide one unit. Proposers are chosen at random. It is easy to see that in any stationary equilibrium, the proposer takes 3/5, offering 1/5 to two others to form a minimum winning coalition.[2]

FGM also study a variant of this model in which one player (an "Apex player") is always chosen with "three times the power". The easiest way to think of the Apex is to imagine him as three players rolled into 1. So there are effectively 7 votes, of which the Apex has 3, and the Apex is also given the floor thrice as often as any other player. Winning coalitions with the Apex included only need contain two players; otherwise they must contain 4. This has, of course, the effect of making the Baron–Ferejohn model generate even more unequal payoffs: for instance, if the Apex proposes, he gets 6/7 and offers a single other player 1/7. It also has the effect of excluding the Apex player very often when other players propose.

The strong and robust finding in their experimental study is that divisions are far more equal than these predictions would suggest. The Apex player is generally included (over 70% of the time) when a winning coalition is formed. Because such a coalition need not contain many people, there is more surplus to go around. This

---

[2]As already noted, the inequality of payoffs arises from the fact that a rejector only has a 1/5 probability of being the new proposer, and that she may be excluded from the winning coalition otherwise.

would not help much in the Baron–Ferejohn theory because the Apex player eats up a lot of that surplus. Here, by contrast, the surplus *does* go around. Therefore surplus-division within the winning coalition is much more equal and also the Apex is included more often than the Baron–Ferejohn model would predict.

FGM run an interesting variant in which the Apex is actually paid (by the experimenters) one-third of what the Apex receives from other players. One interpretation is that the Apex is a coalition of three only one of whom is playing the bargaining game; the remaining two have to be paid off. This changes none of the theoretical predictions — after all, this just amounts to a proportional tax on one of the players and should not change any of the equilibria. But what is interesting is that the experiments now come into much closer line with the theory as far as the inclusion of the Apex is concerned. Other players invite the Apex in only 39% of the time (as opposed to what they were doing before; over 70%).

The broad conclusion from all this is that when proposers form coalitions, they either feel obliged to divide things within the coalition pretty equally, or if such unilateral sentiments are missing, the responders enforce equal division by threatening to reject otherwise.[3] When this is very expensive to do (as in the variant above), they would rather exclude someone altogether. Exclusion appears to be a more bearable instrument than within-group inequality.

14.3 Coalition Formation With Deliberate Exit

We've made a distinction between "final" commitments that are irreversible and "intermediate" commitments that can be reversed, provided the affected parties agree to do so. For instance, within the broad compass of the bargaining approach to coalition formation, this was the main distinction between the two lines of argument initiated in Chapters 4 and 8. But one can clearly entertain extensions of these ideas.

One particularly important line of reasoning considers the case in which a player (or coalition) can *choose* to make one sort of commitment or the other. If it were to choose the irreversible option,

---

[3]The rejector-proposes protocol, which occupies center-stage in much of this book, may be interpreted as a shorthand for generating such within-coalition equity, without having to resort to "other-regarding" preferences.

one might describe it as "exit". Otherwise, it forms provisional agreements which are reversible, and "stays in the game."

It should be noted, however, that a model in which a player can choose a particular commitment structure is not a more "general" model than one in which she cannot, just as a model in which all commitments can be ultimately reversed is in no sense a more "general" model than one in which they cannot. (Or *vice versa*, I might add.) It all depends on the available "technology" that underlies commitment structures, an issue that I have addressed in some detail elsewhere (see, for instance, Section 2.6 in Chapter 2).

Perry and Reny (1994), Seidmann and Winter (1998) and Bloch and Gomes (2006) analyze different models of deliberate exit. Here is a version based on the Bloch-Gomes paper, which uses unlimited upfront transfers in coalition formation. There are an infinite number of dates, as in the reversible commitments model introduced in Section 8, and each date has two stages. At the first *proposal stage*, coalitions are formed. A proposal by i to a set of players S is simply an offer to "buy" out the players in S; if accepted, the entire set can be viewed as a single player/coalition in the following round of proposals.

On the heels of every proposal stage comes an *action stage*, in which each of the going player/coalitions can either choose to irreversibly exit the game by choosing some irreversible action, or choose a reversible action which is fixed for the remainder of that period. One-period discounted payoffs are added together, as are any upfront transfers made during a proposal stage, to arrive at total payoffs.

The Bloch-Gomes analysis suggests that the results are in line with those obtained in Chapters 9 and 10 for binding agreements under reversible commitments. Specifically, when the payoffs of exiting players are unaffected by the actions of those who choose to remain, the long-run outcome of their game is necessarily efficient. Characteristic functions fall into this category, of course, because the actions of one group of players never affects the payoffs of a separate group. But the Bloch-Gomes category is broader than that: it allows for externalities across players as long as they "remain" in the game.

So the results run parallel to those in Chapter 9 but apply to a wider class of games.[4]

On the other hand, there is scope for persistently inefficient outcomes when exit options for a coalition do depend on the actions of other individuals. The discussion in this part of the paper nicely ties up with the case of externalities and reversible commitments, studied in Chapter 10.

I do not put forward the Bloch-Gomes analysis as a fully definitive exercise. For instance, as they themselves discuss, some of their results may vary with the proposer protocol (see, for instance, their discussion of Seidmann and Winter (1998) in Example 3 of their paper). Yet their analysis (along with those of the other authors cited here) take the theory in an interesting discussion, one that is well worth further investigation.

14.4 Overlapping Coalitions

There are many situations in which players may belong to more than one coalition. Firm A and firm B might cooperate in the production of a particular product or service, and so might firms B and C, while firms A and C may have nothing to do with each other. Likewise, free trade agreements are routinely signed across overlapping collections of countries. Or overlapping groups of individuals may be involved in relationships involving reciprocity, information-sharing, or public-goods provision.

Now, none of this is of any serious concern as long as the writing of one such agreement has no effect at all on the worth or value of another agreement. If the formation of coalition S leaves the worths of all other coalitions unchanged, including the worths of those groups that intersect with S, one can go ahead and simply treat each of these as separate bargaining problems. That would be the end of the story.

But of course, matters are generally more complicated. The worths of a formed coalition *do* affect those of another, and they do so in two fundamental ways.

---

[4]A fuller comparison will require more careful study: the results in Chapter 9 hold for all (benign) equilibria, while in the Bloch-Gomes paper attention is restricted to the traditional class of Markovian equilibria.

First, for reasons of law, custom or information, the formation of one coalition may simply negate the formation of some other coalitions. The logic of a partitional coalition structure implicitly presumes that when coalition S forms, *no other coalition that overlaps with it can also form*. In a single-product oligopoly, it makes no sense to say that firm A cooperates with firm B, and B with C, but that A and B interact noncooperatively with each other! The coalition structure $\{AB, BC\}$ simply does not make sense. Likewise, in a customs union (as opposed to a free trade area), the presumption is that there is free flow of factors and products within the union. Two overlapping customs unions don't make sense. Or consider communities that are formed through the provision of local public goods, nonexcludable as long as you are living in the same region.

These examples are all constructed to defend the partitional structure, but the general point at the start of the last paragraph is obviously broader. There are many instances in which a similar constraint holds, but in a nonpartitional way. For instance, the formation of S may rule out separate agreements within *subsets* of S.

The partitional coalition structure — whether or not a characteristic function or a partition function is defined on it — embodies such restrictions. In this sense even the characteristic function carries with it a cross-coalitional externality!

Second, even if the formation of S does not have any bearing on the formation on the ability of another coalition T, it might nonetheless affect what T can achieve. If S and T are overlapping sets of countries, a free-trade agreement within S does not preclude another for T, but the *payoffs* associated with the latter agreement will surely be affected by the former (and *vice versa*). Indeed, we are now on the same familiar turf as the partition function with intercoalitional externalities, and there is little to add.

Some more formalization may be useful here. Let N be a set of players. A *cover* of N is a collection of coalitions $\gamma = \{S_1, \ldots, S_m\}$ such that their union is N. (A partitional coalition structure is a special case of a cover.) A *cover function* assigns to each cover γ a value $v(S, \gamma)$ to every coalition $S \in \gamma$.

Now, some covers (such as nonpartitional covers in partitional situations) may simply be infeasible. This is easy enough to represent provided that the game is normalized to assure positive payoffs for all coalitions that do "legitimately" form. Simply set the values of coalitions in all "illegitimate" covers equal to zero. With this convention in place, it is easy enough to see that both partition functions and characteristic functions are special cases.

This monograph does not handle cover functions. This is not because I foresee serious difficulties in doing so.[5] The text reports on finished work, and cover functions represent part of the road ahead. In my view an extension of the analysis to handle such functions would be very welcome. It may well lead to new theoretical insights, and it will certainly broaden the applicability of the theory developed here.

14.5 Networks

If we've already gone from partitional structures to cover functions, we can go a step further, though what I am about to discuss is — along different dimensions — both more general and more specific.

In several situations, the structure of interactions between individuals is perhaps best described by a *network*. Examples include the sharing of information (Bala and Goyal (2000), Calvo-Armengol and Jackson (2001), Bramoullé and Kranton (2002) and Kariv(2002)), trading networks (Tesfatsion (1997, 1998) and Weisbuch, Kirman and Herreiner (2000)), mutual insurance (Fafchamps and Lund (1997) and Bloch, Genicot and Ray (2007)), technology adoption (Conley and Udry (2002), Chatterjee and Xu (2004) and Bandiera and Rasul (2006)) and buyer–seller networks (Kranton and Minehart (2000, 2001) and Wang and Watts (2006)). This is a burgeoning literature, and Matt Jackson's forthcoming book (Jackson (2007)) will serve as an excellent introduction to it.

The network formalism is used as follows. Describe the space of interactions as a graph, where nodes are just the players, while an arc between two nodes indicates the existence of a "bilateral interaction" between the two corresponding players. Thus a network is just a

---

[5]This is provided we start from cover functions as a primitive. I must note that there are interesting and deep conceptual issues in the derivation of cover functions from an underlying game in strategic form. In particular, it is difficult to apply a simultaneous-move framework, such as the coalitional equilibrium concept in Chapter 3, to noncooperative interactions across coalitions in an overlapping cover.

graph g on N: a collection of ij pairs, the interpretation being that i and j are "linked".[6]

A *component* of a network g is a subset c of g such that no $i \in c$ is linked outside c and such that every distinct i and j in c are directly or indirectly linked. (Thus isolated singletons are components by definition.) For a component c, let $S(c)$ denote the set of individuals in c. Now we can see the sense in which networks generalize coalition structures. The set $S(c)$ may be thought of as a coalition, but it is also an object with finer structure "within" it, for all its members are not directly linked to one another. Formally, we could embed a coalition structure into a graph by simply requiring that every component be *completely connected*: every pair of agents within each component is linked. Indeed, with a bit more work, covers can be integrated within the network formalism as well.

So in this sense, networks are indubitably more general. Yet it is unclear whether an *integrated* theory of coalitions and networks is possible or even desirable, though obviously each literature can gain from the other in terms of methods, existing results, etc. The reason is that network formation places the emphasis — as it should, in my view — squarely on *bilateral* link formation. The presence of other links may well impose an externality on the deliberating pair, but nevertheless it is the pair that decides to form the link, and — in many economic and social contexts — no one else. This view is central to the vast majority of network-formation models starting from the work of Aumann and Myerson (1988).[7] I subscribe to this approach to networks. Yet it is also worth noting that this special feature is also what keeps the networks literature logically separate from coalition formation, in which a group of agents (and not just a pair) may deliberately choose to form. While it is mathematically possible to describe models of link formation that generalize both the bilateral (networks) view and the multilateral (coalitions) view, the significance of such a research agenda from a social-science angle is not entirely evident to me.

---

[6]Because the graph is undirected, these links are reciprocal. For analyses of valuation structures which are directed graphs, see Bala and Goyal(2000) and Dutta and Jackson (2000).

[7]Indeed, there are several situations in which link *breakage* may be an entirely individual choice, even while link *formation* is bilateral. See Jackson and Wolinsky (1996) for a formalization of this viewpoint. See also Dutta and Mutuswami (1997), Dutta, van den Nouweland and Tijs (1998), and Slikker and van den Nouweland (2000, 2001).

That said, there may be much to be gained by applying methods and results from one of these fields to the other. As an example of this, consider the problem of farsighted network formation, studied in Dutta, Ghosal and Ray (2005), henceforth DGR. Suppose that we are interested in the dynamic evolution of a network, with individuals receiving payoffs at every date from the network in place at that date. Then link formation (or breakage) has repercussions not just today, but also at future dates, because the evolution of fresh links will generally depend on the link structure already in place. If individuals understand these repercussions, they will take them into account when forming or breaking a link in the current period.

Suppose, then, that each network g is associated with a set of one-period payoffs, one for each agent. A process of network formation would then simply be a (possibly stochastic) sequence of such networks, one for each date, with each network connected to its immediate predecessor via a set of newly formed or broken links. Such a process would require every individual to follow a certain strategy, one that prescribes the breaking or creation of links, or perhaps inaction, at every stage at which she is supposed to move.

Such a process constitutes an *farsighted equilibrium* if every player sticks to her prescribed strategy, and in so doing evaluates payoffs not just from the current actions but from the entire process that results, now and in the future. In particular, a potentially profitable deviation is not necessarily myopic: individuals take the ongoing process as given and evaluate the entire stream of consequences arising from a single action. One can imitate perfectly myopic behavior by taking the discount factor to zero, and perfect farsightedness by taking the opposite limit. DGR provide the needed formal account.

In this solution concept, network formation and payoffs occur together. In particular, the definition permits cycles and continued flux in the network, without creating any difficulty in the evaluation of overall payoffs. This sort of structure is obviously and closely related to several of the papers on coalition formation discussed in the book,[8] though at the same time the underlying network structure imbues the concept with its own distinctive features (more on this below).

---

[8]The concept is most closely related to Konishi and Ray (2003), Gomes and Jehiel (2005), and Hyndman and Ray (2007).

DGR use this solution concept to study the question of efficiency in dynamic network formation. It is well-known from the work of Jackson and Wolinsky (1996) and others that "stable" networks in a static context may not be efficient. When a link is formed, or destroyed, the players involved do so with their own gain in mind. At the same time, these actions also affect the payoff of other players, and so a wedge is driven between stability and efficiency. A solution concept that incorporates farsightedness allows us to investigate the same questions for dynamic processes of network formation. Does the possibility of constant renegotiation of links remove the static inefficiency, or heighten it? DGR establish a dynamic counterpart of the stability-efficiency tradeoff: there are situations in which the process will not converge to any efficient network for *any* equilibrium strategy profile. But it is also possible to identify situations in which the presence of farsightedness removes inefficiency.

Indeed, there are several other questions one can ask of dynamic networks. Do such processes invariably converge or are they forever in flux? What is the precise connection between the degree of farsightedness (proxied by the discount factor) and the extent of efficiency? What if upfront transfers can be made to encourage link formation? These are interesting issues indeed, but getting into them would be taking us too far afield of the specific focus in this book. Instead, allow me to hope that I've piqued your interest sufficiently for you to think more about such matters.

14.6 Coalition Formation With Nonbinding Agreements

Nonbinding play sets us squarely in the world of noncooperative theory, in which issues of coalition formation are often relegated to a secondary role. It must be presumed, of course, that even when a coalition forms its members must play some noncooperative equilibrium among themselves. A coalition in this world has an indicative role to play, by placing attention on a particular subset of equilibrium strategies, one that is presumably beneficial to the coalition concerned. In short, and presuming that no form of binding play is possible, a coalition can be viewed purely as a device for equilibrium selection.

An early instance of this is the notion of *strong equilibrium*, a Nash equilibrium in which no set of players can gain by moving together to a fresh profile of strategies, while presuming that the complementary set of players continue to abide by the original profile (Aumann (1959)). This refines the set of Nash equilibria, while bestowing powerful stability properties on the equilibria that do survive the criterion: imagine something that is immune not just to individual deviations but *arbitrary* coalitional deviations: that would be both Nash equilibrium and nonempty core rolled into one! But this very conflation of two disparate ideas rings a warning bell. There is no reason to judge a *nonbinding* agreement suspect if it can be blocked by an agreement which itself may lack any stability properties (as a nonbinding agreement in its own right). Thus Bernheim, Peleg and Whinston (1987) introduce the concept of *coalition-proof Nash equilibrium* (CPNE), in which coalitional deviations meet the same rigorous stability criteria that constrain the arrangements that are being deviated upon.

In a sentence, a potential coalitional deviation from a coalition-proof Nash equilibrium must itself pass coalition-proofness in an "induced game" in which the complementary coalition stick to their presumed strategies. This creates a recursive definition.

There is a parallel between such a recursion and the notion of farsightedness in this book. Deviating coalitions "look ahead" not just to a deviation but to the consequences of that deviation. The deviation must therefore be constrained as well (see an initial discussion in Section 2.4 of Chapter 2).

The major difference is that CPNE applies only to games where *no* binding agreements are possible (CPNE are always Nash). In contrast, the solution concepts in this book are emphatically not Nash equilibria of the underlying games they study, though of course they *are* "noncooperative" equilibria of the underlying negotiation process. As a trivial example, the unique noncooperative outcome in the Prisoner's Dilemma is also the unique CPNE, while the only equilibrium binding agreement for the induced partition function is the cooperative outcome.

It is important to understand what lies at the formal heart of this distinction. It is captured by the specification of what follows a coalitional deviation. Imagine two players at a negotiating table. If no binding agreements are possible, the CPE are precisely those Nash equilibria which are Pareto optimal in the class of all Nash equilibria. If binding agreements are possible, the word "deviation" translates as "breaking-off of negotiations". In this example, the environment then shifts to the coalition structure of the two players acting independently (two singleton coalitions). When one of the two players "deviates", there is no question of taking the other player's strategy as given, as in CPE. Therefore, it is imperative to explicitly model the "game" that results after the deviation, and use the "equilibria" of this game to determine the limits of the original negotiation process.

It would be of interest to investigate dynamic noncooperative games with (nonbinding) coalition formation, along lines parallel to those studied in Chapters 8–10 and Chapter 13. One can adopt several approaches to this problem. For instance, one can extend the standard theory of infinitely repeated games (or more generally, dynamic games with a state variable) to allow for an intertemporal analogue of CPNE.[9] This is a worthwhile project but continues to regard coalition formation as no more than an equilibrium selection device. Alternatively, one might begin with the partition function, so that the formation of a coalition structure at any date has a definite impact on payoffs, perhaps through the writing of binding agreements within coalitions in any period. But the important difference from the model of Chapter 8 is that such agreements would — by assumption — be up for grabs at the end of every period. There are no binding agreements that last for longer than a single date. In this sense, the analysis would have more in common with Chapter 13, though specific results for partition functions are yet to be obtained and fully understood.

14.7 Incomplete Information

This book assumes complete information throughout. When a group of agents gets together, it is commonly known what each agent brings to the coalition. Yet there are obviously several economic and political situations in which coalitions form with incomplete information about member types. In some cases, incomplete information isn't the "first-order problem" at hand. For instance, for each of the applications mentioned in this book there is a corresponding variant with incomplete information. But there are other cases in which incomplete information may be the fundamental problem:

---

[9]For an exercise along these lines, see, e.g., Chung (2004).

think, for instance, of cartels in auctions or group formation for the purpose of informal insurance.

Coalitional considerations make an appearance in a variety of situations that involve incomplete information.[10] But it is probably fair to say that an explicit theory of coalition formation in such a context doesn't exist at this time. There is a small but important literature that attempts to develop notions of blocking and the core in an environment of incomplete information; see, for instance, Wilson (1978), Holmström and Myerson (1983), Myerson (1984) and (1991, Chapter 10), and Forges, Mertens and Vohra (2002). The survey by Forges, Minelli and Vohra (2002) contains additional references to the literature and serves as an extremely useful introduction to the subject.

There is another area, quite apart from the one emphasized in this literature, where incomplete information plays a fundamentally important role. This is in the matter of the so-called compensation principle. Consider the problem of building a dam, or the imposition (or removal) of a protective tariff, or the implementation of a new environmental regulation. Uppermost on my mind as I write these pages is the political crisis in the state of West Bengal in India, as the government attempts to acquire agricultural land for industrial purposes. In each of these situations there are losers and gainers, and coalitions will form around the proposed policy. Let us suppose that in each case, the policy does increase aggregate surplus, so that in principle the winners can compensate the losers. Yet how do we identify a loser, and how do we identify a winner? For instance, it might seem trivial to identify "losers" in the land acquisition example: surely these are just the owners of the land? The answer is in the negative: the losing coalition includes sharecroppers and landless laborers, all of whom will be affected by the proposed policy.

The problem of incomplete information is a major stumbling block in the design of proper compensation. In this sense it has fundamental implications for the theory of coalition formation.

---

[10]A good example is mechanism design, in which it is sometimes required that implementation be immune to coalitional deviations.

14.8 Nontransferable Utility

The inability to make transfers or compensatory payments, perhaps due to incomplete information or to the lack of commitment, pushes us to study games with nontransferable utilities. This book certainly doesn't shy away from NTU games, and they make an appearance at several different points. However, I haven't chosen to particularly emphasize the NTU case in the same way that I have emphasized the case of transferable payoffs.

The contribution that is most significantly relevant to the material in this book is Bloch (1996). This paper studies partition functions with fully nontransferable utility. In other words, to each partition of the player set into a coalition structure is associated a *single* payoff division, taken to be completely symmetric within each coalition. Bloch shows how an equilibrium coalition structure is generated in this model.

Bloch's results are closely related to the analysis of symmetric partition functions with *transferable* payoffs, studied in Chapter 5. The reason is simple: we show in that chapter that coalitional bargaining leads to approximately equal division within any coalition that forms, when discount factors are close enough to unity. The forces that govern equilibrium coalition structure are, therefore, roughly the same.

Of course, this is no longer the case in heterogeneous models (see Chapter 7). The analysis in that chapter again predicts a specific payoff allocation in equilibrium when utilities are transferable, but this allocation is not given by equal division within coalitions. Therefore a suitable algorithm appears to be extremely hard for the general NTU case, though possibly not out of reach. Banerjee (2002) makes some progress by providing conditions for the uniqueness of no-delay stationary equilibria (an analogue of Proposition 7.1) in NTU characteristic functions.

The literature on networks is another area in which specific payoff divisions are assumed once a set of links has formed (see, e.g., Jackson and Wolinsky (1996)). A third area in which nontransferable utilities form a leading case is that of "hedonic games", in which each player obtains a deterministic, immutable payoff from the coalition that she chooses to join (Banerjee, Konishi and Sönmez (2001)).

14.9 Axiomatic Approaches

The theory in this book relies explicitly on behavioral models of coalition formation and payoff allocation. One might adopt an alternative viewpoint based on the *axiomatic properties* that a possible solution must satisfy. Such an approach is presumably complementary to a behavioral model. One might be interested in an axiomatic characterization of our equilibria, uncluttered by behavioral detail. Alternatively, one might study game forms that "implement" a solution concept derived from axiomatic first principles.

Ever since von Neumann and Morgenstern characterized expected utility from a primitive set of desiderata, the use of axioms has permeated deeply into decision theory, as well as into cooperative game theory. In cooperative game theory, the approach takes the following specific form. A *solution* is a mapping from a particular domain (characteristic functions, partition functions, bargaining problems, and so on) to a set of outcomes (payoffs, coalition structure, surplus division, etc.). On such a solution or mapping, place a set of axioms, which usually restricts the way in which the solution changes across specific elements of the domain. The strength of the axiomatic method is that it often generates — via these seemingly innocuous restrictions — a *particular* solution. Two celebrated instances of the axiomatic approach in cooperative game theory are the Nash bargaining solution (Nash (1950)) and the Shapley value (Shapley (1953)).

A recent attempt to axiomatically derive "equilibrium" payoff allocations and coalition structures for partition function games is Maskin (2003). Maskin begins with the class of all transferable-utility, superadditive[11] partition functions. The domain of analysis is a pair: a partition function from this class, together with a coalition substructure for that partition function, the latter to be interpreted as a set of coalitions that have "already formed". A solution is a mapping on this domain with *two* outcomes, the first specifying a payoff allocation, and the second describing the "final" coalition structure that "forms".

---

[11]As we have noted at several points, superadditivity is a restrictive assumption. However, it should be noted that this paper represents a first step towards an ambitious axiomatic approach.

Thus a solution runs parallel to the completion maps studied in Section 7.6.2 of Chapter 7, and first introduced in Ray and Vohra (1999, Section 4.1.2). The difference is that a behavioral model is used to generate the completion map, while a set of axioms will be used here to derive a class of solutions.

Maskin (2003) employs four axioms. The first requires that coalitional worth not be wasted: the payoff allocation must be such that the sum of payoffs for each coalition equals the worth of that coalition. The second axiom requires that each player be assigned to a coalition where her marginal contribution is highest. The third requires that she be indeed allocated this marginal contribution. The fourth is a consistency axiom, requiring that as the domain substructure is partially augmented along the lines dictated by the solution, the continuation prescribed by the mapping be unchanged. These axioms are compatible with at least one and at most a finite number of possible solutions (Theorem 1). In the special subclass of characteristic functions, the solution reduces to the Shapley value (Theorem 3).

I do not want to comment in much detail on the axioms, except to note that the axioms on marginal contributions do push the solution towards a not unexpected extension of the Shapley value. But that is precisely the job of an axiomatic study: the researcher is looking to implement a particular solution, and the axioms are the conversation that take place between the researcher and the reader, serving as supportive and presumably persuasive evidence for that solution.[12]

Maskin's theory is most closely related to the general approach espoused in this book, in that it takes explicit note of the possibility that efficient outcomes may not result, and that in superadditive games an equilibrium structure of *sub*coalitions cannot be ruled out. However, there is also a literature that attempts to extend efficient Shapley or Shapley-like solutions to partition functions with externalities. Myerson (1977) initiates this literature; for more recent contributions, see, e.g., Bolger (1989) and Macho-Stadler, Pérez-Castrillo and Wettstein (2007). Owen (1977) extends the Shapley value to coalition structures. McQuillin (2006) represents a recent attempt to unite the Myerson and Owen approaches. In its simultaneous consideration of both externalities and coalition

---

[12]In his Toulouse lectures (2004), available as slides on the web but not, to my knowledge, in paper form, Maskin revisits these axioms, unpacking them somewhat more in an attempt to characterize the same set of solutions.

structure, this last paper is probably closest to Maskin's, but stresses (as do the other cited papers) the desideratum of efficiency as a normative requirement for a solution, which Maskin does not.

References

ABREU, D. (1988), "Towards a Theory of Discounted Repeated Games," *Econometrica* **56**, 383–396.

AGHION, P., ANTRAS, P. and E. HELPMAN (2004), "Negotiating Free Trade," mimeograph, Department of Economics, Harvard University.

ANDREONI, J. and J. MILLER (2002), "Giving According to GARP: An Experimental Test of the Consistency of Preferences for Altruism", *Econometrica* **40**, 737–753.

ASH, R. (1972), *Real Analysis and Probability*. Academic Press, Orlando.

AUMANN, R. (1959), "Acceptable Points in General Cooperative *n*-Person Games," in *Contributions to the Theory of Games IV*, Annals of Mathematics Study 40, (A. Tucker and R. Luce (eds.)), Princeton, NJ: Princeton University Press.

AUMANN, R. (1961), "The Core of a Cooperative Game Without Sidepayments," *Transactions of the American Mathematical Society* **98**, 539–552.

AUMANN, R. and M. MASCHLER (1964), "The Bargaining Set for Cooperative Games," in *Advances in Game Theory* (M. Dresher, L. Shapley, and A. Tucker, eds.), Annals of Mathematical Studies No. 52; Princeton, NJ: Princeton University Press.

AUMANN, R., and R. MYERSON (1988), "Endogenous Formation of Links Between Players and of Coalitions: An Application of the Shapley Value," in *The Shapley Value: Essays in Honor of Lloyd Shapley*, A. Roth, ed., . 175–191. Cambridge: Cambridge University Press.

BALA, V. and S. GOYAL (2001) "Self-Organization in Communication Networks," *Econometrica* **68**, 1181–1230.

BANDIERA, O. and I. RASUL (2006), "Social Networks and Technology Adoption in Northern Mozambique," *Economic Journal* **116**, 869–902.

BANERJEE, S. (2002) "A Theory of Coalitional Bargaining in a Non-Transferable Utility Game," mimeograph, Department of Economics, Boston University.

BANERJEE, S., KONISHI, H. and T. SÖNMEZ (2001), "Core in a Simple Coalition Formation Game, *Social Choice and Welfare* **18**, 135–153.

BARON, D. AND J. FEREJOHN (1989), "Bargaining in Legislatures," *American Political Science Review* **83**, 1181–1206.

BERNHEIM, D., PELEG, B. and M. WHINSTON (1987), "Coalition-Proof Nash Equilibria. I. Concepts," *Journal of Economic Theory* **42**, 1–12.

BINMORE, K. (1985), "Bargaining and Coalitions," in *Game-Theoretic Models of Bargaining* (A. Roth (ed.)), Cambridge: Cambridge University Press.

BLOCH, F. (1996), "Sequential Formation of Coalitions in Games with Externalities and Fixed Payoff Division," *Games and Economic Behavior* **14**, 90–123.

BLOCH, F. and A. GOMES (2006), "Contracting with Externalities and Outside Options," *Journal of Economic Theory* **127**, 172–201.

BLOCH, F., SOUBEYRAN, R. and S. SÁNCHEZ-PAGÉS (2006), "When does Universal Peace Prevail? Secession and Group Formation in Conflict," *Economics of Governance* **7**, 3–29.

BLOCH F., GENICOT, G. and D. RAY (2007), "Informal Insurance in Social Networks," mimeograph, Department of Economics, New York University.

BOLGER, E. (1989), "A Set of Axioms for a Value for Partition Function Games," *International Journal of Game Theory* **18**, 37–44.

BRAMOULLÉ, Y. and R. KRANTON (2002), "Social Learning, Social Networks, and Search," mimeograph, Department of Economics, University of Maryland.

CALVO-ARMENGOL, A. and M. JACKSON (2001), "Social Networks and the Resulting Dynamics and Patterns of Employment and Wages," mimeograph, CalTech.

CARRARO, C. and D. SINISCALCO (1993), "Strategies for the International Protection of the Environment, *Journal of Public Economics* **52**, 309–328.

CHANDER, P. and H. TULKENS (1995), The Core of an Economy with Multilateral Environmental Externalities," CORE Discussion Paper No. 9550.

CHARNESS, G. and B. GROSSKOPF (2001), "Relative Payoffs and Happiness: An Experimental Study," *Journal of Economic Behavior and Organization* **45**, 301–328.

CHARNESS, G. and M. RABIN (2002), "Understanding Social Preferences With Simple Tests," *Quarterly Journal of Economics* **117**(3), 817–869.

CHATTERJEE, K., DUTTA, B., RAY, D. and K. SENGUPTA (1993), "A Nonco-operative Theory of Coalitional Bargaining," *Review of Economic Studies* **60**, 463–477.

CHATTERJEE, K. and S. XU (2004), "Technology Diffusion by Learning from Neighbors," *Advances in Applied Probability* **36**, 355–376.

CHUNG, A. (2004), "Coalition-Stable Equilibria in Repeated Games," mimeograph, Department of Economics, Stanford University.

CHWE, M. (1994), "Farsighted Coalitional Stability," *Journal of Economic Theory* **63**, 299–325.

COASE, R. (1960), "The Problem of Social Cost," *The Journal of Law and Economics* **3**, 1–44.

CONLEY, T. and C. UDRY (2002), "Learning About a New Technology: Pineapple in Ghana," mimeograph, Department of Economics, Yale University.

DIAMANTOUDI, E. and XUE, L. (2007), "Coalitions, Agreements and Efficiency," *Journal of Economic Theory* **136**, 105–125.

DUBEY, P. (1986), "Inefficiency of Nash Equilibria," *Mathematics of Operations Research* **11**, 1–8.

DUTTA, B. and M. JACKSON (2000), "The Stability and Efficiency of Directed Communication Networks," *Review of Economic Design* **5**, 251–272.

DUTTA, B., and S. MUTUSWAMI (1997) "Stable Networks," *Journal of Economic Theory* **76**, 322–344.

DUTTA, B., VAN DEN NOUWELAND, A., and S. TIJS (1998) "Link Formation in Cooperative Situations," *International Journal of Game Theory* **27**, 245–256.

DUTTA, B. and D. RAY (1989), "A Concept of Egalitarianism Under Participation Constraints," *Econometrica* **57**, 615–635.

DUTTA, B. and D. RAY (1991), "Constrained Egalitarian Allocations," *Games and Economic Behavior* **3**, 403–422.

DUTTA, B., RAY, D., SENGUPTA, K. and R. VOHRA (1989), "A Consistent Bargaining Set, *Journal of Economic Theory* **49**, 93–112.

DUTTA, B. and K. SUZUMURA (1993), "On the Sustainability of R&D Through Private Incentives," Indian Statistical Institute Discussion Paper No. 93-13.

DUTTA, B., GHOSAL, S. and D. RAY, "Farsighted Network Formation," *Journal of Economic Theory* **122**, 143–164.

ESTEBAN, J. and RAY, D. (1999), "Conflict and Distribution", *Journal of Economic Theory* **87**, 379–415.

ESTEBAN, J. and J. SÁKOVICS (2004), "Olson vs. Coase: Coalitional Worth in Conflict," *Theory and Decision* **55**, 339–357.

FAFCHAMPS, M. and S. LUND (2000) "Risk-Sharing Networks in Rural Philippines," mimeograph, Stanford University.

FORGES, F., MERTENS, J-F. and R. VOHRA (2002), "The Ex Ante Incentive Compatible Core in the Absence of Wealth Effects," *Econometrica* **70**, 1865–1892.

FRECHETTE, G., KAGEL, J. and M. MORELLI (2005), "Behavioral Identification in Coalitional Bargaining: An Experimental Analysis of Demand Bargaining and Alternating Offers," *Econometrica* **73**, 1893–1938.

FUDENBERG, D. and E. MASKIN (1986), "The Folk Theorem in Repeated Games with Discounting or with Incomplete Information," *Econometrica* **54**, 533–556.

GOMES, A. and P. JEHIEL (2005), "Dynamic Processes of Social and Economic Interactions: On the Persistence of Inefficiencies," *Journal of Political Economy* **113**, 626–667.

GREEN, J. (1974), "The Stability of Edgeworth's Recontracting Process," *Econometrica* **42**, 21–34.

GREENBERG, J. (1990), *The Theory of Social Situations*, Cambridge, MA: Cambridge University Press.

HAERINGER, G. (2004), "Equilibrium Binding Agreements: A comment," *Journal of Economic Theory* **117**, 140–143.

HARRIS, C. (1985), "Existence and Characterization of Perfect Equilibrium in Games of Perfect Information," *Econometrica* **53**, 613–628.

HARSANYI, J. (1963), "A Simplified Bargaining Model for the *n*-Person Cooperative Game," *International Economic Review* **4**, 194–220.

HARSANYI, J. (1974), "An Equilibrium-Point Interpretation of Stable Sets and a Proposed Alternative Definition," *Management Science* **20**, 1472–1495.

HARSANYI, J. and R. SELTEN (1988), *A General Theory of Equilibrium Selection in Games*, Cambridge, MA and London: MIT Press.

HART, S. and M. KURZ (1983), "Endogenous Formation of Coalitions," *Econometrica* **51**, 1047–1064.

HERRERO, M. J. (1985), "*n*-player Bargaining and Involuntary Unemployment," Ph.D. Dissertation, London School of Economics.

HOLMSTRÖM, B. and R. MYERSON (1983), "Efficient and Durable Decision Rules with Incomplete Information," *Econometrica* **51**, 1799–1819.

HYNDMAN, K. and D. RAY (2007), "Coalition Formation with Binding Agreements," forthcoming, *Review of Economic Studies*.

ICHIISHI, T. (1981), "A Social Coalitional Equilibrium Existence Lemma," *Econometrica* **49**, 369–37.

Jackson, M. (2007), *Social and Economic Networks*, forthcoming, Princeton University Press.

Jackson, M. and A. Wolinsky (1996), "A Strategic Model of Social and Economic Networks," *Journal of Economic Theory*, 71, 44–74.

Kalandrakis, A. (2004), "A Three-Player Dynamic Majoritarian Bargaining Game," *Journal of Economic Theory* 116, 294–322.

Kandori, M., Mailath, G. and R. Rob (1993), "Learning, Mutation, and Long Run Equilibria in Games," *Econometrica* 61, 29–56.

Kariv, S. (2002), "Social Learning in a Network," mimeograph, Department of Economics, New York University.

Kawamori, T. (2006), "Noncooperative Coalitional Bargaining with Generalized Selection of Proposers," mimeograph, Graduate School of Economics, University of Tokyo.

Konishi, H. and D. Ray (2003), "Coalition Formation as a Dynamic Process," *Journal of Economic Theory* 110, 1–41.

Kranton, R. and D. Minehart (2000), "Competition for Goods in Buyer–Seller Networks," *Review of Economic Design* 5, 301–332.

Kranton, R. and D. Minehart (2001), "A Theory of Buyer–Seller Networks," *American Economic Review*.

Krishna, P. (1998), "Regionalism vs Multilateralism: A Political Economy Approach," *Quarterly Journal of Economics* 113, 227–250.

Krugman, P. (1993), "Regionalism versus Multilateralism: Analytical Notes," in *New Dimensions in Regional Integration* (J. de Melo and A. Panagariya (eds.)), Cambridge, UK: Cambridge University Press.

Lagunoff, R. and A. Matsui (1997), "Asynchronous Choice in Repeated Coordination Games," *Econometrica* 65, 1467–1477.

Levy, G. (2004), "A Model of Political Parties," *Journal of Economic Theory* 115, 250–277.

Lucas, W. (1963), "On Solutions to *n*-Person Games in Partition Function Form," Ph.D. dissertation, University of Michigan, Ann Arbor.

Lucas, W. (1968), "A Game with No Solution," *Bulletin of the American Mathematical Society* 74, 237–239.

Macho-Stadler, I., Pérez-Castrillo, D. and D. Wettstein (2007), "Sharing the Surplus: An Extension of the Shapley value for Environments with Externalities," *Journal of Economic Theory* 135, 339–356.

Mariotti, M. (1997), "A Model of Agreements in Strategic Form Games," *Journal of Economic Theory* 74, 196–217.

Mas-Colell, A. (1989), "An Equivalence Theorem for a Bargaining Set," *Journal of Mathematical Economics* 18, 129–138.

MASKIN, E. (2003), "Bargaining, Coalitions and Externalities," Presidential address of the Econometric Society.

McQUILLIN, B. (2006), "The Extended and Generalized Shapley Value: Simultaneous Consideration of Coalitional Externalities and Coalitional Structure," mimeograph, School of Economics, University of East Anglia.

MOLDOVANU, B. (1992), "Coalition-Proof Nash Equilibria and the Core in Three-Player Games," *Games and Economic Behavior* **4**, 565–581.

MYERSON, R. (1977), "Values for Games in Partition Function Form," *International Journal of Game Theory* **6**, 23–31.

MYERSON, R. (1984), "Cooperative Games with Incomplete Information," *International Journal of Game Theory* **13**, 69–96.

MYERSON, R. (1991), *Game Theory: Analysis of Conflict*, Cambridge, MA: Harvard University Press.

NASH, J. (1950), "The Bargaining Problem," *Econometrica* **18**, 155–162.

OKADA, A. (1996), "A Noncooperative Coalitional Bargaining Game with Random Proposers," *Games and Economic Behavior* **16**, 97–108.

OKADA, A. (2000), "The Efficiency Principle In Non-Cooperative Coalitional Bargaining", *Japanese Economic Review* **51**, 34–50.

OKADA, A. (2007), "Coalitional Bargaining Games with Random Proposers: Theory and Application," mimeograph, Graduate School of Economics, Hitotsubashi University.

OSBORNE, M. and A. RUBINSTEIN (1994), *A Course in Game Theory*. Cambridge, MA: MIT Press.

OWEN, G. (1977), "Values of Games with A Priori Unions," in *Essays in Mathematical Economics and Game Theory* (R. Henn and O. Moschlin (eds.)), New York, NY: Springer Verlag.

PERRY, M. and P. RENY (1994), "A Non-Cooperative View of Coalition Formation and the Core," *Econometrica* **62**, 795–817.

PHELPS, E. AND R. POLLAK (1968), "On Second-Best National Saving and Game Equilibrium Growth," *Review of Economic Studies* **35**, 185–199.

RAY, D. (1989), "Credible Coalitions and the Core," *International Journal of Game Theory* **18**, 185–187.

RAY, D. (1998), *Development Economics*, Princeton, NJ: Princeton University Press.

RAY, D. and R. VOHRA (1997), "Equilibrium Binding Agreements," *Journal of Economic Theory.* **73**(1), 30–78.

RAY, D. and R. VOHRA (1999), "A Theory of Endogenous Coalition Structures," *Games and Economic Behavior* **26**, 286–336.

RAY, D. and R. VOHRA (2001), "Coalitional Power and Public Goods," *Journal of Political Economy* **109**, 1355–1384.

RIKER, W. (1962), *The Theory of Political Coalitions*, New Haven, CT: Yale University Press.

ROSENTHAL, R. (1972), "Cooperative Games in Effectiveness Form," *Journal of Economic Theory* **5**, 88–101.

RUBINSTEIN, A. (1982), "Perfect Equilibrium in a Bargaining Model," *Econometrica* **50**, 97–109.

SALANT, S., SWITZER, S. and R. REYNOLDS (1983), "Losses from Horizontal Merger: The Effects of an Exogenous Change in Industry Structure on Cournot–Nash Equilibrium," *Quarterly Journal of Economics* **93**, 185–199.

SCARF, H. (1971), "On the Existence of a Cooperative Solution for a General Class of *n*-Person Games," *Journal of Economic Theory* **3**, 169–181.

SEIDMANN, D. (2005), "Preferential Trading Arrangements as Strategic Positioning," mimeograph, School of Economics, University of Nottingham.

SEIDMANN, D. and E. WINTER (1998), "Gradual Coalition Formation," *Review of Economic Studies* **65**, 793–815.

SELTEN, R. (1981), "A Non-Cooperative Model of Characteristic Function Bargaining," in *Essays in Game Theory and Mathematical Economics in Honor of Oscar Morgenstern* (V. Böhm and H. Nachtkamp (eds.)), Mannheim: Bibliographisches Institut.

SENGUPTA, A. and K. SENGUPTA (1996), "A Property of the Core," *Games and Economic Behavior* **12**, 266–73.

SHAFER, W. and H. SONNENSCHEIN (1975), "Equilibrium in Abstract Econo- mies Without Ordered Preferences," *Journal of Mathematical Economics* **2**, 345–348.

SHAPLEY, L. (1953), "A Value for *n*-Person Games," in *Contributions to the Theory of Games, II.* (H. Kuhn and A. Tucker (eds.)), Princeton, NJ: Princeton University Press.

SHENOY, P. (1979), "On Coalition Formation: A Game Theoretic Approach," *International Journal of Game Theory* **8**, 133–164.

SHUBIK, M. (1983), *Game Theory in the Social Sciences*, Cambridge, MA: MIT Press.

SLIKKER, M. and A. VAN DEN NOUWELAND (2000), "Network Formation Models with Costs for Establishing Links," *Review of Economic Design* **5**, 333–362.

SLIKKER, M. and A. VAN DEN NOUWELAND (2001), "A One-stage Model of Link Formation and Payoff Division", *Games and Economic Behavior* **34**, 153–175.

STÅHL, I. (1977), "An *n*-Person Bargaining Game in the Extensive Form," in *Mathematical Economics and Game Theory* (R. Henn and O. Moeschlin (eds.)) Berlin: Springer-Verlag.

STROTZ, R. (1956), "Myopia and Inconsistency in Dynamic Utility Maximization," *Review of Economic Studies* **23**, 165–180.

SUTTON, J. (1986) "Non-Cooperative Bargaining Theory: An Introduction," *Review of Economic Studies* **53**, 709–724.

TESFATSION, L. (1997), "A Trade Network Game with Endogenous Partner Selection," in H. Amman et al (eds.), *Computational Approaches to Economic Problems*, Dordrecht: Kluwer Academic Publishers, 249–269.

TESFATSION, L. (1998), "Gale–Shapley Matching in an Evolutionary Trade Network Game," Iowa State University Economic Report no. 43.

THRALL, R. and W. LUCAS (1963), "*n*-Person Games in Partition Function Form," *Naval Research Logistics Quarterly* **10**, 281–298.

VAN DAMME, E., SELTEN, R. and E. WINTER (1990), "Alternating Bid Bargaining with a Smallest Money Unit," *Games and Economic Behavior* **2**, 188–201.

VON NEUMANN, J. and O. MORGENSTERN (1944), *Theory of Games and Economic Behavior*, Princeton, NJ: Princeton University Press.

WANG, P. and A. WATTS (2006), "Formation of Buyer–Seller Trade Networks in a Quality-Differentiated Product Market," *Canadian Journal of Economics* **39**, 971–1004.

WEISBUCH, G., KIRMAN, A. and D. HERREINER (2000), "Market Organisation and Trading Relationships," *Economic Journal* **110**, 411–436.

WILSON, R. (1978), "Information, Efficiency and the Core of an Economy," *Econometrica* **46**, 807–816.

WINTER, E. (1993), "Mechanism Robustness in Multilateral Bargaining," *Theory and Decision* **40**, 131–47.

XUE, L. (1998), "Coalitional Stability under Perfect Foresight," *Economic Theory* **11**, 603–627.

YI, S-S. (1996), "Endogenous Formation of Customs Unions under Imperfect Competition: Open Regionalism is Good," *Journal of International Economics* **41**, 153–177.

YOUNG, P. (1993), "The Evolution of Conventions," *Econometrica* **61**, 57–84.

ZHAO, J. (1992), "The Hybrid Solutions of an *n*-Person Game," *Games and Economic Behavior* **4**, 145–160.

Subject Index

Author Index